Deceitful Media

Deceitful Media

Artificial Intelligence and Social Life after the Turing Test

Simone Natale

Oxford University Press is a department of the University of Oxford. It furthers the University's objective of excellence in research, scholarship, and education by publishing worldwide. Oxford is a registered trade mark of Oxford University Press in the UK and certain other countries.

Published in the United States of America by Oxford University Press
198 Madison Avenue, New York, NY 10016, United States of America.

Library of Congress Cataloging-in-Publication Data
Names: Natale, Simone, 1981– author.
Title: Deceitful media : artificial intelligence and social life after the Turing test /
Simone Natale.
Description: New York : Oxford University Press, [2021] |
Includes bibliographical references and index.
Identifiers: LCCN 2020039479 (print) | LCCN 2020039480 (ebook) |
ISBN 9780190080365 (hardback) | ISBN 9780190080372 (paperback) |
ISBN 9780190080396 (epub)
Subjects: LCSH: Artificial intelligence—Social aspects. | Philosophy of mind.
Classification: LCC Q335 .N374 2021 (print) | LCC Q335 (ebook) |
DDC 303.48/34—dc23
LC record available at https://lccn.loc.gov/2020039479
LC ebook record available at https://lccn.loc.gov/2020039480

DOI: 10.1093/oso/9780190080365.001.0001

CONTENTS

Acknowledgments *vii*

Introduction *1*

1. The Turing Test: The Cultural Life of an Idea *16*
2. How to Dispel Magic: Computers, Interfaces, and the Problem of the Observer *33*
3. The ELIZA Effect: Joseph Weizenbaum and the Emergence of Chatbots *50*
4. Of Daemons, Dogs, and Trees: Situating Artificial Intelligence in Software *68*
5. How to Create a Bot: Programming Deception at the Loebner Prize Competition *87*
6. To Believe in Siri: A Critical Analysis of Voice Assistants *107*

Conclusion: Our Sophisticated Selves *127*

Notes *133*
Bibliography *161*
Index *187*

ACKNOWLEDGMENTS

When I started working on this book, I had an idea about a science fiction story. I might never write it, so I reckon it is just fine to give up its plot here. A woman, Ellen, is awakened by a phone call. It's her husband. There is something strange in his voice; he sounds worried and somehow out of tune. In the close future in which this story is set, artificial intelligence (AI) has become so efficient that a virtual assistant can make calls on your behalf by reproducing your own voice, and the simulation will be so accurate as to trick even your close family and friends. Ellen and her husband, however, have agreed that they would never use AI to communicate between them. Yet in the husband's voice that morning there is something that doesn't sound like him. Later, Ellen discovers that her husband has died that very night, a few hours before the time of their call. The call should have been made by an AI assistant. Dismayed by her loss, she listens to the conversation again and again until she finally picks up some hints to solve the mystery. In fact, this science fiction story I haven't written is also a crime story. To learn the truth about her husband's death, Ellen will need to interpret the content of the conversation. In the process, she will also have to establish whether the words came from her husband's, from the machine that imitated him, or from some combination of the two.

This book is not science fiction, yet like much science fiction, it is also an attempt to make sense of technologies whose implications and meaning we are just starting to understand. I use the history of AI—a surprisingly long one for technologies that are often presented as absolute novelties—as a compass to orient my exploration. I started working on this book in 2016. My initial idea was to write a cultural history of the Turing test, but my explorations brought exciting and unexpected discoveries that made the final project expand much beyond that.

A number of persons read and commented on early drafts of this work. My editor, Sarah Humphreville, not only believed in this project since the start but also provided crucial advice and punctual suggestions throughout its development. Assistant Editor Emma Hodgon was also exceedingly helpful and scrupulous. Leah Henrickson provided feedback on all the chapters; her intelligence and knowledge made this just a much better book. I am grateful to all who dedicated time and attention to read and comment on different parts of this work: Saul Albert, Gabriele Balbi, Andrea Ballatore, Paolo Bory, Riccardo Fassone, Andrea Guzman, Vincenzo Idone Cassone, Nicoletta Leonardi, Jonathan Lessard, Peppino Ortoleva, Benjamin Peters, Michael Pettit, Thais Sardá, Rein Sikveland, and Cristian Vaccari.

My colleagues at Loughborough University have been a constant source of support, both professionally and personally, during the book's gestation. I would like especially to thank John Downey for being such a generous mentor at an important and potentially complicated moment of my career, and for teaching me the importance of modesty and integrity in the process. Many other senior staff members at Loughborough were very supportive in many occasions throughout the last few years, and I wish particularly to thank Emily Keightley, Sabina Mihelj, and James Stanyer for their constant help and friendliness. Thanks also to my colleagues and friends Pawas Bisht, Andrew Chadwick, David Deacon, Antonios Kyparissiadis, Line Nyhagen, Alena Pfoser, Marco Pino, Jessica Robles, Paula Saukko, Michael Skey, Elzabeth Stokoe, Vaclav Stetka, Thomas Thurnell-Read, Peter Yeandle, and Dominic Wring, as well as to all other colleagues at Loughborough, for making work easier and more enjoyable.

During the latest stages of this project I was awarded a Visiting Fellowship at ZeMKI, the Center for Media, Communication, and Information Research of the University of Bremen. It was a great opportunity to discuss my work and to have space and time to reflect and write. Conversations with Andreas Hepp and Yannis Theocharis were particularly helpful to clarify and deepen some of my ideas. I thank all the ZeMKI members for their feedback and friendship, especially but not only Stefanie Averbeck-Lietz, Hendrik Kühn, Kerstin Radde-Antweiler, and Stephanie Seul, as well as the other ZeMKI Fellows whose residence coincided with my stay: Peter Lunt, Ghislain Thibault, and Samuel Van Ransbeeck.

Some portions of this book have been revised from previous publications. In particular, parts of chapter 3 were previously published, in a significantly different version, in the journal *New Media and Society*, and an earlier version of chapter 6 was featured as a Working Paper in the *Communicative*

Figurations Working Papers series. I thank the reviewers and editors for their generous feedback.

My thanks, finally, go to the many humans who acted as my companions throughout these years, doing it so well that no machine will ever be able to replace them. This book is especially dedicated to three of them: my brother and sister and my partner, Viola.

I remember a visit I made to one of Queen Victoria's residences, Osborne on the Isle of Wight. . . . Prominent among the works displayed there was a life-size marble sculpture of a large furry dog, a portrait of the Queen's beloved pet "Noble." The portrait must have been as faithful as the dog undoubtedly was—but for the lack of color it might have been stuffed. I do not know what impelled me to ask our guide, "May I stroke him?" She answered, "Funny you want to do that; all the visitors who pass stroke him—we have to wash him every week." Now, I do not think the visitors to Osborne, myself included, are particularly prone to magic beliefs. We did not think the image was real. But if we had not thought it somewhere we would hardly have reacted as we did—that stroking gesture may well have been compounded of irony, playfulness, and a secret wish to reassure ourselves that after all the dog was only of marble.

—Ernst Gombrich, *Art and Illusion*

Introduction

In May 2018, Google gave a public demonstration of its ongoing project Duplex, an extension of Google Assistant programmed to carry out phone conversations. Google's CEO, Sundar Pichai, presented the recording of a conversation in which the program mimicked a human voice to book an appointment with a hair salon. Duplex's synthetic voice featured pauses and hesitation in an effort to sound more credible. The strategy appeared to work: the salon representative believed she was speaking with a real person and accepted the reservation.[1]

In the following weeks, Duplex's apparent achievements attracted praise, but also criticism. Commentaries following the demonstration highlighted two problems about the demo. On one side, some contended that Duplex operated "straight up, deliberate deception,"[2] opening new ethical questions regarding the capacity of an artificial intelligence (AI) to trick users into believing it is human. On the other side, some expressed doubts about the authenticity of the demo. They pointed to a series of oddities in the recorded conversations: the businesses, for instance, never identified themselves, no background noise could be heard, and the reservation-takers never asked Duplex for a contact number. This suggested that Google might have doctored the demo, faking Duplex's capacity to pass as human.[3]

Deceitful Media. Simone Natale, Oxford University Press (2021). © Oxford University Press.
DOI: 10.1093/oso/9780190080365.003.0001

The controversy surrounding Duplex reflects a well-established dynamic in the public debate about AI. Since its inception in the 1950s, the achievements of AI have often been discussed in binary terms: either exceptional powers are attributed to it, or it is dismissed as a delusion and a fraud.[4] Time after time, the gulf between these contradictory assessments has jeopardized our capacity to recognize that the true impact of AI is more nuanced and oblique than usually acknowledged. The same risk is present today, as commentators appear to believe that the question should be whether or not Duplex is able to pass as a human. However, even if Google's gadget proved unable to pass as human, we should not believe the illusion to be dispelled. Even in the absence of deliberate misrepresentation, AI technologies entail forms of deception that are perhaps less evident and straightforward but deeply impact societies. We should regard deception not just as a possible way to employ AI but as a constitutive element of these technologies. Deception is as central to AI's functioning as the circuits, software, and data that make it run.

This book argues that, since the beginning of the computer age, researchers and developers have explored the ways users are led to believe that computers are intelligent. Examining the historical trajectory of AI from its origins to the present day, I show that AI scientists have incorporated knowledge about users into their efforts to build meaningful and effective interactions between humans and machines. I call, therefore, for a recalibration of the relationship between deception and AI that critically questions the ways computing technologies draw on specific aspects of users' perception and psychology in order to create the illusion of AI.

One of the foundational texts for AI research, Alan Turing's *Computing Machinery and Intelligence* (1950), set up deception as a likely outcome of interactions between humans and intelligent computers. In his proposal for what is now commonly known as the Turing test, he suggested evaluating computers on the basis of their capacities to deceive human judges into believing they were human. Although tricking humans was never the main objective of AI, computer scientists adopted Turing's intuition that whenever communication with humans is involved, the behavior of the human users informs the meaning and impact of AI just as much as the behavior of the machine itself. As new interactive systems that enhanced communications between humans and computers were introduced, AI scientists began more seriously engaging with questions of how humans react to seemingly intelligent machines. The way this dynamic is now embedded in the development of contemporary AI voice assistants such as Google Assistant, Amazon's Alexa, and Apple's Siri signals the emergence of a new kind of interface, which mobilizes deception in order to manage the interactions between users, computing systems, and Internet-based services.

Since Turing's field-defining proposal, AI has coalesced into a disciplinary field within cognitive science and computer science, producing an impressive range of technologies that are now in public use, from machine translation to the processing of natural language, and from computer vision to the interpretation of medical images. Researchers in this field nurtured the dream—cherished by some scientists while dismissed as unrealistic by others—of reaching "strong" AI, that is, a form of machine intelligence that would be practically indistinguishable from human intelligence. Yet, while debates have largely focused on the possibility that the pursuit of strong AI would lead to forms of consciousness similar or alternative to that of humans, where we have landed might more accurately be described as the creation of a range of technologies that provide an *illusion* of intelligence—in other words, the creation not of intelligent beings but of technologies that humans perceive as intelligent.

Reflecting broader evolutionary patterns of narratives about technological change, the history of AI and computing has until now been mainly discussed in terms of technological capability.[5] Even today, the proliferation of new communicative AI systems is mostly explained as a technical innovation sparked by the rise of neural networks and deep learning.[6] While approaches to the emergence of AI usually emphasize evolution in programming and computing technologies, this study focuses on how the development of AI has also built on knowledge about users.[7] Taking up this point of view helps one to realize the extent to which tendencies to project agency and humanity onto things makes AI potentially disruptive for social relations and everyday life in contemporary societies. This book, therefore, reformulates the debate on AI on the basis of a new assumption: that what machines are changing is primarily us, humans. "Intelligent" machines might one day revolutionize life; they are already transforming how we understand and carry out social interactions.

Since AI's emergence as a new field of research, many of its leading researchers have professed to believe that humans are fundamentally similar to machines and, consequently, that it is possible to create a computer that equals or surpasses human intelligence in all aspects and areas. Yet entertaining a similar tenet does not forcefully contrast with and is often complementary to the idea that existing AI systems provide only the illusion of human intelligence. Throughout the history of AI, many have acknowledged the limitations of present systems and focused their efforts on designing programs that would provide at least the appearance of intelligence; in their view, "real" or "strong" AI would come through further progress, with their own simulation systems representing just a step in that direction.[8] Understanding how humans engage in social exchanges, and how they can be led to treat things as social agents, became instrumental to overcoming the limitations of AI technologies. Researchers in AI thus established a direction of research that was based on the designing of technologies that cleverly exploited human perception and expectations to give users the impression of employing or interacting with intelligent systems. This book demonstrates that looking at the development across time of this tradition—which has not yet been studied as such—is essential to understanding contemporary AI systems programmed to engage socially with humans. In order to pursue this agenda, however, the problem of deception and AI needs to be formulated under new terms.

ON HUMANS, MACHINES, AND "BANAL DECEPTION"

When the great art historian Ernst Gombrich started his inquiry into the role of illusion in the history of art, he realized that figurative arts emerge within an interplay between the limits of tradition and the limits of perception. Artists have always incorporated deception into their work, drawing on their knowledge both of convention and of mechanisms of perception to achieve certain effects on the viewer.[9] But who would blame a gifted painter for employing deceit by playing with perspective or depth to make a tableau look more convincing and "real" in the eyes of the observer?

While this is easily accepted from an artist, the idea that a software developer employs knowledge about how users are deceived in order to improve human-computer interaction is likely to encounter concern and criticism. In fact, because the term *deception* is usually associated with malicious endeavors, the AI and computer science communities have proven resistant to discussing their work in terms of deception, or have discussed deception as an unwanted outcome.[10] This book, however, contends that deception is a constitutive element of human-computer interactions rooted in AI technologies. We are, so to say, programmed to be deceived, and modern media have emerged within the spaces opened by the limits and affordances of our capacity to fall into illusion. Despite their resistance to consider deception as such, computer scientists have worked since the early history of their field to exploit the limits and affordances of our perception and intellect.[11]

Deception, in its broad sense, involves the use of signs or representations to convey a false or misleading impression. A wealth of research in areas such as social psychology, philosophy, and sociology has shown that deception is an inescapable fact of social life with a functional role in social interaction and communication.[12] Although situations in which deception is intentional and manifest, such as frauds, scams, and blatant lies, shape popular understandings of deception, scholars have underlined the more disguised, ordinary presence of deception in everyday experience.[13] Many forms of deception are not so clear-cut, and in many cases deception is not even understood as such.[14]

Moving from a phenomenological perspective, philosopher Mark A. Wrathall influentially argued that our capacity to be deceived is an inherent quality of our experience. While deception is commonly understood in binary terms, positing that one might either be or not be deceived, Wrathall contends that such a dichotomy does not account for how people perceive and understand external reality: "it rarely makes sense to say that I perceived either truly or falsely" since the possibility of deception is ingrained in the mechanisms of our perception. If, for instance, I am walking in the woods and believe I see a deer to my side where in fact there is just a bush, I am deceived; yet the same mechanism that made me see a deer where it wasn't—that is, our tendency and ability to identify patterns in visual information—would have helped me, on another occasion, to identify a potential danger. The fact that our senses have shortcomings, Wrathall points out, represents a resource as much as a limit for human perception and is functional to our ability to navigate the external world.[15] From a similar point of view, cognitive psychologist Donald D. Hoffman recently proposed that evolution has shaped our perceptions into useful illusions that help us navigate the physical world but can also be manipulated through technology, advertising, and design.[16]

Indeed, the institutionalization of psychology in the late nineteenth and early twentieth centuries already signaled the discovery that deception and illusion were integral, physiological aspects of the psychology of perception.[17] Understanding deception was important not much or not only in order to study how people misunderstood the world but also to study how they perceived and navigated it.[18] During the nineteenth and twentieth centuries, the accumulation of knowledge about how people were deceived informed the development of a wide range of media technologies and practices, whose effectiveness exploited the affordances and limitations of our senses of seeing, hearing, and touching.[19] As I demonstrate in this book, AI developers, in order to produce their outcomes, have continued this tradition of technologies that mobilize our liability to deception. Artificial intelligence scientists have collected information and knowledge about how users react to machines that exhibit the appearance of intelligent behaviors, incorporating this knowledge into the design of software and machines.

One potential objection to this approach is that it dissolves the very concept of deception by equating it with "normal" perception. I contend, however, that rejecting a binary understanding of deception helps one realize that deception involves a wide spectrum of situations that have very different outcomes but also common characteristics. If on one end of the spectrum there are explicit attempts to mislead, commit fraud, and tell lies, on the other end there are forms of deception that are not so clear-cut and that, in many cases, are not understood as such.[20] Only by identifying and studying less evident dynamics of deception can we develop a full understanding of more evident and straight-out instances of deception. In pointing to the centrality of deception, therefore, I do not intend to suggest that all forms of AI have hypnotic or manipulative goals. My main goal is not to establish whether AI is "good" or "bad" but to explore a crucial dimension of AI and interrogate how we should proceed in response to this.

Home robots such as Jibo and companion chatbots such as Replika, for example, are designed to appear cute and to awaken sentiments of empathy in their owners. This design choice looks in itself harmless and benevolent: these technologies simply work better if their appearance and behavior stimulate positive feelings in their users.[21] The same characteristics, however, will appear less innocent if the companies producing these systems start profiting from these feelings in order to influence users' political opinions. Home robots and companion chatbots, together with a wide range of AI technologies programmed to enter into communication with humans, structurally incorporate forms of deception: elements such as appearance, a humanlike voice, and the use of specific language expressions are designed to produce specific effects in the user. What makes this less or more acceptable is not the question whether there is or is not deception but what the outcomes and the implications are of the deceptive effects produced by any given AI technology. Broadening the definition of deception, in this sense, can lead to improving our comprehension of the potential risks of AI and related technologies, counteracting the power of the companies that gain from the user's interactions with these technologies and stimulating broader investigations of whether such interactions pose any potential harm to the user.

To distinguish from straight-out and deliberate deception, I propose the concept of *banal deception* to describe deceptive mechanisms and practices that are embedded in media technologies and contribute to their integration into everyday life. Banal deception entails mundane, everyday situations in which technologies and devices mobilize specific elements of the user's perception and psychology—for instance, in the case of AI, the all-too-human tendency to attribute agency to things or personality to voices. The word "banal" describes things that are dismissed as ordinary and unimportant; my use of this word aims to underline that these mechanisms are often taken for granted, despite their significant impact on the uses and appropriations of media technologies, and are deeply embedded in everyday, "ordinary" life.[22]

Different from approaches to deliberate or straight-out deception, banal deception does not understand users and audiences as passive or naïve. On the contrary, audiences actively exploit their own capacity to fall into deception in sophisticated ways—for example, through the entertainment they enjoy when they fall into the illusions offered by cinema or television. The same mechanism resonates with the case of AI. Studies in human-computer interaction consistently show that users interacting with computers apply norms and behaviors that they would adopt with humans, even if these users perfectly understand the difference between computers and humans.[23] At first glance, this seems incongruous, as if users resist and embrace deception simultaneously. The concept of banal deception provides a resolution of this apparent contradiction. I argue that the subtle dynamics of banal deception allow users to embrace deception so that they can better incorporate AI into their everyday lives, making AI more meaningful and useful to them. This does not mean that banal deception is harmless or innocuous. Structures of power often reside in mundane, ordinary things, and banal deception may finally bear deeper consequences for societies than the most manifest and evident attempts to deceive.

Throughout this book, I identify and highlight five key characteristics that distinguish banal deception. The first is its everyday and ordinary character. When researching people's perceptions of AI voice assistants, Andrea Guzman was surprised by what she sensed was a discontinuity between the usual representations of AI and the responses of her interviewees.[24] Artificial intelligence is usually conceived and discussed as extraordinary: a dream or a nightmare that awakens metaphysical questions and challenges the very definition of what means to be human.[25] Yet when Guzman approached users of systems such as Siri, the AI voice assistant embedded in iPhones and other Apple devices, she did not find that the users were questioning the boundaries between humans and machines.

Instead, participants were reflecting on themes similar to those that also characterize other media technologies. They were asking whether using the AI assistant made them lazy, or whether it was rude to talk on the phone in the presence of others. As Guzman observes, "neither the technology nor its impact on the self from the perspective of users seemed extraordinary; rather, the self in relation to talking AI seemed, well, ordinary—just like any other technology."[26] This ordinary character of AI is what makes banal deception so imperceptible but at the same time so consequential. It is what prepares for the integration of AI technologies into the fabrics of everyday experience and, as such, into the very core of our identities and selves.[27]

The second characteristic of banal deception is functionality. Banal deception always has some potential value to the user. Human-computer interaction has regularly employed representations and metaphors to build reassuring and easily comprehensible systems, hiding the complexity of the computing system behind the interface.[28] As noted by Michael Black, "manipulating user perception of software systems by strategically misrepresenting their internal operations is often key to producing compelling cultural experiences through software."[29] Using the same logic, communicative AI systems mobilize deception to achieve meaningful effects. The fact that users behave socially when engaging with AI voice assistants, for instance, has an array of pragmatic benefits: it makes it easier for users to integrate these tools into domestic environments and everyday lives, and presents possibilities for playful interaction and emotional reward.[30] Being deceived, in this context, is to be seen not as a misinterpretation by the user but as a response to specific affordances coded into the technology itself.

The third characteristic of banal deception is obliviousness: the fact that the deception is not understood as such but taken for granted and unquestioned. The concept of "mindless behavior" has been already used to explain the apparent contradiction, mentioned earlier, of AI users understanding that machines are not human but still to some extent treating them as such.[31] Researchers have drawn from cognitive psychology to describe mindlessness as "an overreliance on categories and distinctions drawn in the past and in which the individual is context-dependent and, as such, is oblivious to novel (or simply alternative) aspects of the situation."[32] The problem with this approach is that it implies a rigid distinction between mindfulness and mindlessness whereby only the latter leads to deception. When users interact with AI, however, they also replicate social behaviors and habits in self-conscious and reflective ways. For instance, users carry out playful exchanges with AI voice assistants, although they know too well the machine will not really get their jokes. They wish them goodnight before going to bed, even if aware that they will not "sleep" in the same sense as humans do.[33] This suggests that distinctions between mindful and mindless behaviors fail to capture the complexity of the interaction. In contrast, obliviousness implies that while users do not thematize deception as such, they may engage in social interactions with the machine deliberately as well as unconsciously. Obliviousness also allows the user to maintain at least the illusion of control—this being, in the age of user-friendliness, a key principle of software design.[34]

The fourth characteristic of banal deception is its *low definition*. While this term is commonly used to describe formats of video or sound reproduction with lower resolution, in media theory the term has also been employed in reference to media that demand more participation from audiences and users in the construction of sense and meaning.[35] For what concerns AI, textual and voice interfaces are low definition because they leave ample space for the user to imagine and attribute characteristics such as gender, race, class, and personality to the disembodied voice or text. For instance, voice assistants do not present at a physical or visual level the appearance of the virtual character (such as "Alexa" or "Siri"), but some cues are embedded in the sounds of their voices, in their names, and in the content of their exchanges. It is for this reason that, as shown in research about people's perceptions of AI voice assistants, different users imagine AI assistants in different, multiple ways, which also enhances the effect of technology being personalized to each individual.[36] In contrast, humanoid robots leave less space for the users' imagination and projection mechanisms and are therefore not low definition. This is one of the reasons why disembodied AI voice assistants have become much more influential today than humanoid robots: the fact that users can project their own imaginations and meanings makes interactions with these tools much more personal and reassuring, and therefore they are easier to incorporate into our everyday lives than robots.[37]

The fifth and final defining characteristic of banal deception is that it is not just imposed on users but also is programmed by designers and developers. This is why the word *deception* is preferable to *illusion*, since *deception* implies some form of agency, permitting clearer acknowledgment of the ways developers of AI technologies work toward achieving the desired effects. In order to explore and develop the mechanisms of banal deception, designers need to construct a model or image of the expected user. In actor-network theory, this corresponds to the notion of script, which refers to the work of innovators as "inscribing" visions or predictions about the world and the user in the technical content of the new object and technology.[38] Although this is always an exercise of imagination, it draws on specific efforts to gain knowledge about users, or more generally about "humans." Recent work in human-computer interaction acknowledges that "perhaps the most difficult aspect of interacting with humans is the need to model the beliefs, desires, intentions preferences, and expectations of the human and situate the interaction in the context of that model."[39] The historical excavation undertaken in this book shows that this work of modeling users is as old as AI itself. As soon as interactive systems were developed, computer scientists and AI researchers explored how human perception and psychology functioned and attempted to use such knowledge to close the gap between computer and user.[40]

It is important to stress that for us to consider the agency of programmers and developers who design and prepare for use AI systems is perfectly compatible with the recognition that users themselves have agency. As much critical scholarship on digital media shows, in fact, users of digital technologies and systems often subvert and reframe the intentions and expectations of companies and developers.[41] This does not imply, however, that the latter do not have an expected outcome in mind. As Taina Bucher recently remarked, "the cultural beliefs and values held by programmers, designers, and creators of software matter": we should examine and question their intentions despite the many difficulties involved in reconstructing them retrospectively from the technology and its operations.[42]

Importantly, the fact that banal deception is not to be seen as negative by default does not mean that its dynamics should not be the subject of attentive critical inquiry. One of the key goals of this book is to identify and counteract potentially problematic practices and implications that emerge as a consequence of the incorporation of banal deception into AI. Unveiling the mechanisms of banal deception, in this sense, is also an invitation to interrogate what the "human" means in the discursive debates and practical work that shape the development of AI. As the trajectory described in this book demonstrates, the modeling of the "human" that has been developed throughout the history of AI has in fact been quite limited. Even as computer access has progressively been extended to wider potential publics, developers have often envisioned the expected user as a white, educated man, perpetuating biases that remain inherent in contemporary computer systems.[43] Furthermore, studies and assumptions about how users perceive and react to specific representations of gender, race, and class have been implemented in interface design, leading for instance to gendered characterizations of many contemporary AI voice assistants.[44]

One further issue is the extent to which the mechanisms of banal deception embedded in AI are changing the social conventions and habits that regulate our relationships with both humans and machines. Pierre Bourdieu uses the concept of habitus to characterize the range of dispositions through which individuals perceive and react to the social world.[45] Since habitus is based on previous experiences, the availability of increasing opportunities to engage in interactions with computers and AI is likely to feed forward into our social behaviors in the future. The title of this book refers to AI and social life *after* the Turing test, but even if a computer program able to pass that test is yet to be created, the dynamics of banal deception in AI already represent an inescapable influence on the social life of millions of people around the world. The main objective of this book is to neutralize the opacity of banal deception, bringing its mechanisms to the surface so as to better understand new AI systems that are altering societies and everyday life.

ARTIFICIAL INTELLIGENCE, COMMUNICATION, MEDIA HISTORY

Artificial intelligence is a highly interdisciplinary field, characterized by a range of different approaches, theories, and methods. Some AI-based applications, such as the information-processing algorithms that regulate access to the web, are a constant presence in the everyday lives of masses of people; others, like industrial applications of AI in factories and workshops, are rarely, if ever, encountered.[46] This book focuses particularly on communicative AI, that is, AI applications that are designed to enter into communication with human users.[47] Communicative AIs include applications involving conversation and speech, such as natural language processing, chatbots, social media bots, and AI voice assistants. The field of robotics makes use of some of the same technologies developed for communicative AI—for instance to have robots communicate through a speech dialogue system—but remains outside the remit of this book. As Andreas Hepp has recently pointed out, in fact, AI is less commonly in use today in the form of embodied physical artifacts than software applications.[48] This circumstance, as mentioned earlier, may be explained by the fact that computers do not match one of the key characteristics of banal deception: low definition.

Communicative AI departs from the historical role of media as mere channels of communication, since AI also acts as a producer of communication, with which humans (as well as other machines) exchange messages.[49] Yet communicative AI is still a medium of communication, and therefore inherits many of the dynamics and structures that have characterized mediated communication at least since the emergence of electronic media in the nineteenth century. This is why, to understand new technologies such as AI voice assistants or chatbots, it is vital to contextualize them in the history of media.

As communication technologies, media draw from human psychology and perception, and it is possible to look at media history in terms of how deceitful effects were incorporated in different media technologies. Cinema achieves its effects by exploiting the limits of human perception, such as the impression of movement that can be given through the fast succession of a series of still images.[50] Similarly, as Jonathan Sterne has aptly shown, the development of sound media drew from knowledge about the physical and psychological characteristics of human hearing and listening.[51] In this sense, the key event of media history since the nineteenth century was not the invention of any new technology such as the telegraph, photography, cinema, television, or the computer. It was instead the emergence of the new human sciences, from physiology and psychology to the social sciences, that provided the knowledge and epistemological framework to adapt modern media to the characteristics of the human sensorium and intellect.

Yet the study of media has often fallen into the same trap as those who believe that deception in AI matters only if it is "deliberate" and "straight-up."[52] Deception in media history has mainly been examined as an exceptional circumstance, highlighting the manipulative power of media rather than acknowledging deception's structural role in modern media. According to an apocryphal but persistent anecdote, for instance, early movie audiences exchanged representation for reality and panicked before the image of an incoming train.[53] Similarly, in the story of Orson Welles's radio broadcast *War of the Worlds*, which many reportedly interpreted as a report of an actual extraterrestrial invasion, live broadcasting has led people to confuse fiction with reality.[54] While such blatant (and often exaggerated) cases of deception have attracted much attention, few have reflected on the fact that deception is a key feature of media technologies' function—that deception, in other words, is not an incidental but an irremediable characteristic of media technologies.[55]

To uncover the antecedents of AI and robotics, historians commonly point to automata, self-operating machines mimicking the behavior and movements of humans and animals.[56] Notable examples in this lineage include the mechanical duck built by French inventor Jacques de Vaucanson in 1739, which displayed the abilities of eating, digesting, and defecating, and the Mechanical Turk, which amazed audiences in Europe and America in the late eighteenth and early nineteenth centuries with its proficiency at playing chess.[57] In considering the relationship between AI and deception, these automata are certainly a case in point, as their apparent intelligence was the result of manipulation by their creators: the mechanical duck had feces stored in its interior, so that no actual digestion took place, while the Turk was maneuvered by a human player hidden inside the machine.[58] I argue, however, that to fully understand the broader relationship between contemporary AI and deception, one needs to delve into a wider historical context that goes beyond the history of automata and programmable machines. This context is the history of deceitful media, that is, of how different media and practices, from painting and theatre to sound recording, television, and cinema, have integrated banal deception as a strategy to achieve particular effects in audiences and users. Following this trajectory shows that some of the dynamics of communicative AI are in a relationship of continuity with the ways audiences and users have projected meaning onto other media and technology.

Examining the history of communicative AI from the proposal of the Turing test in 1950 to the present day, I ground my work in the persuasion that a historical approach to media and technological change helps us comprehend ongoing transformations in the social, cultural, and political spheres. Scholars such as Lisa Gitelman, Erkki Huhtamo, and Jussi Parikka have compellingly shown that what are now called "new media" have a long history, whose study is necessary to understand today's digital culture.[59] If it is true that history is one of the best tools for comprehending the present, I believe that it is also one of the best instruments, although still an imperfect one, for anticipating the future. In areas of rapid development such as AI, it is extremely difficult to forecast even short- and medium-term development, let alone long-term changes.[60] Looking at longer historical trajectories across several decades helps to identify key trends and trajectories of change that have characterized the field across several decades and might, therefore, continue to shape it in the future. Although it is important to understand how recent innovations like neural networks and deep learning work, a better sense is also needed of the directions through which the field has moved across a longer time frame. Media history, in this sense, is a science of the future: it not only sheds light on the dynamics by which we have arrived where we are today but helps pose new questions and problems through which we may navigate the technical and social challenges ahead.[61]

Following Lucy Suchman, I use the terms "interaction" and "communication" interchangeably, since interaction entails the establishment of communication between different entities.[62] Early approaches in human-computer interaction recognized that interaction was always intended as a communicative relationship, and the idea that the computer is both a channel and a producer of communication is much older than often implied.[63] Although AI and human-computer interaction are usually framed as separate, considering them as distinct limits historians' and contemporary communicators' capacity to understand their development and impact. Since the very origins of their field, AI researchers have reflected on how computational devices could enter into contact and dialogue with human users, bringing the problems and questions relevant to human-computer interaction to the center of their own investigation. Exploring the intersections between these fields helps one to understand that they are united by a key tenet: that when a user interacts with technology, the responsibility for the outcome of such interaction is shared between the technology and the human.

On a theoretical level, the book is indebted to insights from different disciplinary fields, from action-network theory to social anthropology, from media theory to film studies and art history. I use these diverse frameworks as tools to propose an approach to AI and digital technologies that emphasizes humans' participation in the construction of meaning. As works in actor-network theory, as well as social anthropologists such as Armin Appadurai and Alfred Gell, have taught us, not only humans but also artifacts can be regarded as social agents in particular social situations.[64] People often attribute intentions to objects and machines: for instance, car owners attribute personalities to their cars and children to their dolls. Things, like people, have social lives, and their meaning is continually negotiated and embedded within social relations.[65]

In media studies, scholars' examinations of the implications of this discovery have shifted from decades-long reflections on the audiences of media such as radio, cinema, and television to the development of a new focus on the interactive relationships between computers and users. In *The Media Equation*, a foundational work published in the mid-1990s, Byron Reeves and Clifford Nass argue that we tend to treat media, including but not only computers, in accordance with the rules of social interaction.[66] Later studies by Nass, Reeves, and other collaborators have established what is known as the Computers Are Social Actors paradigm, which contends that humans apply social rules and expectations to computers, and have explored the implications of new interfaces that talk and listen to users, which are becoming increasingly available in computers, cars, call centers, domestic environments, and toys.[67] Another crucial contribution to such endeavors is that of Sherry Turkle. Across several decades, her research has explored interactions between humans and AI, emphasizing how their relationship does not follow from the fact that computational objects really have emotions or intelligence but from what they evoke in their users.[68]

Although the role of deception is rarely acknowledged in discussions of AI, I argue that interrogating the ethical and cultural implications of such dynamics is an urgent task that needs to be approached through interdisciplinary reflection at the crossroads between computer science, cognitive science, social sciences, and the humanities. While the public debate on the future of AI tends to focus on the hypothesis that AI will make computers as intelligent or even more intelligent than people, we also need to consider the cultural and social consequences of deceitful media providing the appearance of intelligence. In this regard, the contemporary obsession with apocalyptic and futuristic visions of AI, such as the singularity, superintelligence, and the robot apocalypse, makes us less aware of the fact that the most significant implications of AI systems are to be seen not in a distant future but in our ongoing interactions with "intelligent" machines.

Technology is shaped not only by the agency of scientists, designers, entrepreneurs, users, and policy-makers but also by the kinds of questions we ask about it. This book hopes to inspire readers to ask new questions about the relationship between humans and machines in today's world. We will have to start searching for answers ourselves, as the "intelligent" machines we are creating can offer no guidance on such matters-as one of those machines admitted when I asked it (I.1).

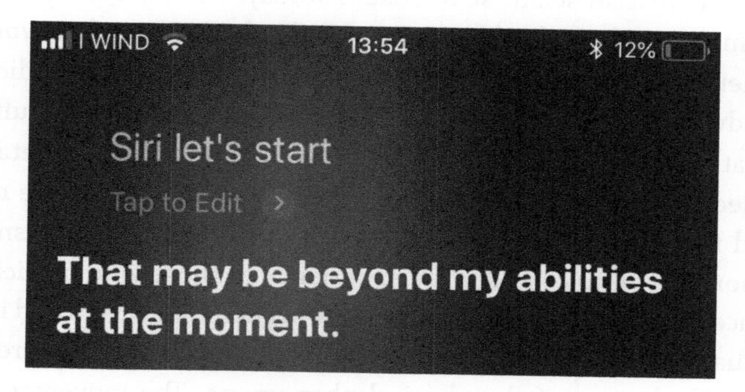

I.1 Author's conversation with Siri, 16 January 2020.

CHAPTER 1
The Turing Test

The Cultural Life of an Idea

In the mid-nineteenth century, tables suddenly took a life of their own. It started in a little town in upstate New York, where two adolescent sisters, Margaret and Kate Fox, reportedly engaged in communication with the spirit of a dead man. Soon word spread, first in the United States and then in other countries, and by 1850 a growing following of people were joining spiritualist séances. The phenomena observed were extraordinary. As a contemporary witness reported, "tables tilt without being touched, and the objects on them remain immobile, contradicting all the laws of physics. The walls tremble, the furniture stamps its feet, candelabra float, unknown voices cry from nowhere—all the phantasmagoria of the invisible populate the real world."[1]

Among those who set out to question the origins of these weird phenomena was British scientist Michael Faraday. His interest was not accidental: many explained table turning and other spiritualist wonders as phenomena of electricity and magnetism, and Faraday had dedicated his life to advancing scientific knowledge in these areas.[2] The results of his investigations, however, would point toward a different interpretation. He contended that the causes of the movements of the tables were not to be searched for in external forces such as electricity and magnetism. It was much more useful to consider the experience of those who participated in the séances. Based on his observations, he argued that sitters and mediums at spiritualist séances not only perceived things that weren't there but also even provoked some of the physical phenomena. The movements of the

Deceitful Media. Simone Natale, Oxford University Press (2021). © Oxford University Press.
DOI: 10.1093/oso/9780190080365.003.0002

séance table could be explained with their unconscious desire for deceiving themselves, which led them involuntarily to move the table.

It wasn't spirits of the dead that made spiritualism, Faraday pointed out. It was the living people—we, the humans—who made it.[3]

Robots and computers are no spirits, yet Faraday's story provides a useful lens through which to consider the emergence of AI from an unusual perspective. In the late 1940s and early 1950s, often described as the gestation period of AI, research in the newly born science of computing placed more and more emphasis on the possibility that electronic digital computers would develop into "thinking machines."[4] Historians of culture, science, and technology have shown how this involved a hypothesis of an equivalence between human brains and computing machines.[5] In this chapter, however, I aim to demonstrate that the emergence of AI also entailed a different kind of discovery. Some of the pioneers in the new field realized that the possibility of thinking machines depended on the perspective of the observers as much as on the functioning of computers. As Faraday in the 1850s suggested that spirits existed in séance sitters' minds, these researchers contemplated the idea that AI resided not so much in circuits and programming techniques as in the ways humans perceive and react to interactions with machines. In other words, they started to envision the possibility that it was primarily users, not computers, that "made" AI.

This awareness did not emerge around a single individual, a single group, or even a single area of research. Yet one contribution in the gestational field of AI is especially relevant: the "Imitation Game," a thought experiment proposed in 1950 by Alan Turing in his article "Computing Machinery and Intelligence." Turing described a game—today more commonly known as the Turing test—in which a human interrogator communicates with both human and computer agents without knowing their identities, and is challenged to distinguish the humans from the machines. A lively debate about the Turing test started shortly after the publication of Turing's article and continues today, through fierce criticisms, enthusiastic approvals, and contrasting interpretations.[6] Rather than entering the debate by providing a unique or privileged reading, my goal here is to illuminate one of the several threads Turing's idea opened for the nascent AI field. I argue that, just as Faraday placed the role of humans at the center of the question of what happened in spiritualist séances, Turing proposed to define AI in terms of the perspective of human users in their interactions with computers. The Turing test will thus serve here as a theoretical lens more than as historical evidence—a reflective space that invites us to interrogate the past, the present, and the future of AI from a different standpoint.

IS THERE ANYTHING LIKE A THINKING MACHINE?

Artificial intelligence sprang up in the middle of the twentieth century at the junction of cybernetics, control theory, operations research, psychology, and the newly born computer science. In this context, researchers exhibited the ambitious goal of integrating these research areas so as to move toward the implementation of human intelligence in general, applicable to any domain of human activity, for example language, vision, and problem-solving. While the foundation of AI is usually dated to 1956, coinciding with a groundbreaking conference at which the term *AI* was introduced, researchers had started to engage with the problem of "machine intelligence" and "thinking machines" at least as early as the early 1940s, coincidentally with the development of the first electronic digital computers.[7] In those early days, as computers were mostly used to perform calculations that required too much time to be completed by humans, imagining that computers would take up tasks such as writing in natural language, composing music, recognize images, or playing chess required quite an imagination. But the pioneers who shaped the emergence of the field did not lack vision. Years before actual hardware and software were created that could handle such tasks, thinkers such as Claude Shannon, Norbert Wiener, and Alan Turing, among others, were confident that it was only a matter of time before such feats would be achieved.[8]

If electronic computers' potential to take up more and more tasks was clear to many, so, however, were the challenges the burgeoning field faced. What was this "intelligence" that was supposed to characterize the new machines? How did it compare to human intelligence? Did it make sense to describe computers' electronic and mechanical processes as "thinking"? And was there any fundamental analogy between the functionings of human brains and of machines that operated by computing numbers? Some, like the two American scientists Warren McCulloch, a neurophysiologist, and Walter Pitts, a logician, believed that human reasoning could be formalized mathematically through logics and calculus, and they formulated a mathematical model of neural activity that laid the ground for successful applications, decades later, of neural networks and deep learning technologies.[9] The view that the brain could be compared and understood as a machine was also shared by the founder of cybernetics, Norbert Wiener.[10]

Yet another potential way to tackle these questions also emerged in this period. It was the result, in a way, of a failure. Many realized that the equation between computers and humans did not stand up to close scrutiny. This was not only due to the technical limitations of the computers of the time. Even if computers could eventually rival or even surpass humans in tasks that were considered to require intelligence, there would still be little evidence to compare their operations to what happens in humans' own minds. The problem can be illustrated quite well by using the argument put forward by philosopher Thomas Nagel in an exceedingly famous article published two decades later, in 1974, titled "What Is It Like to Be a Bat?" Nagel demonstrates that even a precise insight into what happens inside the brain and body of a bat would not make us able to assess whether the bat is conscious. Since consciousness is a subjective experience, one needs to be "inside" the bat to do that.[11] Transferred into the machine intelligence problem, despite computing's tremendous achievements, their "intelligence" would not be similar or equal to that of humans. While all of us has some understanding of what "thinking" is, based on our own subjective experience, we cannot know whether others, and especially non-human beings, share the same experience. We have no objective way, therefore, to know if machines are "thinking."[12]

A trained mathematician who engaged with philosophical issues only as a personal interest, Turing was far from the philosophical sophistication of Nagel's arguments. Turing's objective, after all, was to illustrate the promises of modern computing, not to develop a philosophy of the mind. Yet he showed similar concerns in the opening of "Computer Machinery and Intelligence" (1950). In the introduction, he considers the question "Can machines think?" only to declare it of little use, due to the difficulty of finding agreement about the meanings of the words "machine" and "think." He proposes therefore to replace this question with another one, and thereby introduces the Imitation Game as a more sensible way to approach the issue:

> The new form of the problem can be described in terms of a game which we call the "imitation game." It is played with three people, a man (A), a woman (B), and an interrogator (C) who may be of either sex. The interrogator stays in a room apart from the other two. The object of the game for the interrogator is to determine which of the other two is the man and which is the woman. . . . We now ask the question, "What will happen when a machine takes the part of A in this game?" Will the interrogator decide wrongly as often when the game is played like this as he does when the game is played between a man and a woman? These questions replace our original, "Can machines think?"[13]

As some of his close acquaintances testified, Turing intended the article to be not so much a contribution to philosophy or computer science as a form of propaganda that would stimulate philosophers, mathematicians, and scientists to engage more seriously with the machine intelligence question.[14] But regardless of whether he actually thought of it as such, there is no doubt that the test proved to be an excellent instrument of propaganda for the field. Drawing from the powerful idea that a "computer brain" could beat humans in one of the skills that define our intelligence, the use of natural language, the Turing test presented potential developments of AI in an intuitive and fascinating fashion.[15] In the following decades, it became a staple reference in popular publications presenting the achievements and the potentials of computing. It forced readers and commentators to consider the possibility of AI—even if just rejecting it as science fiction or charlatanry.

THE PLACE OF HUMANS

Turing's article was ambiguous in many respects, which favored the emergence of different views and controversies about the meaning of the Turing test.[16] Yet one of the key implications for the AI field is evident. The question, Turing tells his readers, is not whether machines are or are not able to think. It is, instead, whether we *believe* that machines are able to think—in other words, if we are prepared to accept machines' behavior as intelligent. In this respect, Turing turned around the problem of AI exactly as Faraday did with spiritualism. Much as the Victorian scientist deemed humans and not spirits responsible for producing spiritualist phenomena at séances, the Turing test placed humans rather than machines at the very center of the AI question. Although some have pointed out that the Turing test has "failed" because its dynamic does not accurately adhere to AI's current state of the art,[17] Turing's proposal located the prospects of AI not just in improvements of hardware and software but in a more complex scenario emerging from the interactions between humans and computers. By placing humans at the center of its design, the Turing test provided a context wherein AI technologies could be conceived in terms of their credibility to human users.[18]

There are three actors in the Turing test, all of which engage in acts of communication: a computer player, a human player, and a human evaluator or judge. The computer's capacity to simulate the ways humans behave in a conversation is obviously one of the factors that inform the test's outcome. But since human actors engage actively in communications within the test, their behaviors will be another decisive factor. Things such as their backgrounds, biases, characters, genders, and political opinions will have a role in both the decisions of the interrogator and the behavior of the human agents with which the interrogator interacts. A computer scientist with knowledge and experience on AI, for instance, will be in a different position from someone who has limited insight into the topic. Likewise, human players who act as conversation partners in the test will have their own motivations and ideas on how to participate in the test. Some, for instance, could be fascinated by the possibility of being exchanged for a computer and therefore tempted to create ambiguity about their identities. Because of the role human actors play in the Turing test, all these are variables that may inform its outcome.[19]

The uncertainty that derives from this is often indicated to be one of the test's shortcomings.[20] The test appears entirely logical and justified, however, if one sees it not as an evaluation of the existence of thinking machines but as a measure of humans' reactions to communications with machines that exhibit intelligent behavior. From this point of view, the test made clear for the first time, decades before the emergence of online communities, social media bots, and voice assistants, that AI not only is a matter of computer power and programming techniques but also resides—and perhaps especially—in the perceptions and patterns of interaction through which humans engage with computers.

The wording of Turing's article provides some additional evidence to support this interpretation. Though he refused to make guesses about the question whether machines can think, he did not refrain from speculating that "at the end of the century the use of words and general educated opinion will have altered so much that one will be able to speak of machines thinking without expecting to be contradicted."[21] It is worth reading the wording of this statement attentively. This is not a comment about the development of more functional or more credible machines. It is about cultural change. Turing argues that by the end of the twentieth century people will have a different understanding of the AI problem, so that the prospect of "thinking machines" will not sound as unlikely as it did at Turing's time. He shows interest in "the use of words and general educated opinion" rather than in the possibility of establishing whether machines actually "think."

Looking at the history of computing since Turing's time, there is no doubt that cultural attitudes about computing and AI did change, and quite considerably. The work of Sherry Turkle, who studied people's relationships with technology across several decades, provides strong evidence of this. For instance, in interviews conducted in the late 1970s, as she asked participants what they thought about the idea of chatbots providing psychotherapy services, she encountered much resistance. Most interviewees tended to agree that machines could not make up for the loss of the empathic feeling between the psychologist and the patient.[22] In the following two decades, however, much of this resistance dissolved, with participants in Turkle's later studies becoming more open toward the possibility of psychotherapy chatbots. Discussion of computer psychotherapy has become less moralistic, and her questions were now considered in terms of the limitations on what computers could do or be. People were more willing to concede that, if this was beneficial to the patient, it was worth to try. As Turkle put it, they were "more likely to say, 'Might as well try it. It might help. What's the harm?' "[23]

Turing's prediction might not have been fulfilled—one may still expect today to be contradicted when speaking of machines "thinking"—yet he was right in realizing that cultural attitudes would shift, as a consequence of both the evolution of computing and people's experiences with these technologies. Importantly, because of the role human actors play in the Turing test, such shifts may also inform its outcomes.

Recall the comparison with the spiritualist séance. Sitters joining a "spiritualist circle" witness events such as noises, movements of the table, and the levitation of objects. The outcome of the séance depends on the interpretation the sitters give to these events. It will be informed, as Faraday intuited, not only by the nature of the phenomena but also, and perhaps especially, by the perspective of the sitters. Their attitudes and beliefs, even their psychology and perception, will inform their interpretation.[24] The Turing test tells us that the same dynamics shape interactions between computers and humans. By defining AI in terms of a computer's ability to pass the Turing test, Turing included humans in the equation, making their ideas and biases, as well as their psychology and character, a crucial variable in the construction of "intelligent" machines.

THE COMMUNICATION GAME

Media historian John Durham Peters famously argued that the history of communication can be read as the history of the aspiration to establish an empathic connection with others, and the fear that this may break down.[25] As media such as the telegraph, the telephone, and broadcasting were introduced, they all awoke hopes that they could facilitate such connections, and at the same time fears that the electronic mediation they provided would increase our estrangement from our fellow human beings. It is easy to see how this also applies to digital technologies today. In its relatively short history, the Internet kindled powerful hopes and fears: think, for instance, of the question whether social networks facilitate the creation of new forms of communication or make people lonelier than ever before.[26]

But computing has not always been seen in relationship to communication. In 1950, when Turing published his article, computer-mediated communication had not yet coalesced into a meaningful field of investigation. Computers were mostly discussed as calculating tools, and available forms of interactions between human users and computers were minimal.[27] Imagining communication between humans and computers in 1950, when Turing wrote his article, required a visionary leap—perhaps even greater than that needed to consider the possibility of "machine intelligence."[28] The very idea of a user, understood as an individual given access to shared computing resources, was not fully conceptualized before developments in time-sharing systems and computer networks in the 1960s and the 1970s made computers available to individual access, initially within small communities of researchers and computer scientists and then for an increasingly larger public.[29] As a consequence, the Turing test is usually discussed as a problem about the definition of intelligence. An alternative way to look at it, however, is to consider it as an experiment in the communications between humans and computers. Recently, scholars have started to propose such an approach. Artificial intelligence ethics expert David Gunkel, for instance, points out that "Turing's essay situates communication—and a particular form of deceptive social interaction—as the deciding factor" and should therefore be considered a contribution to computer-mediated communication *avant la lettre*.[30] Similarly, in a thoughtful book written after his experience as a participant in the Loebner Prize—a contemporary contest, discussed at greater length in chapter 5, in which computer programs engage in a version of the Turing test—Brian Christian stresses this aspect, noting that "the Turing test is, at bottom, about the act of communication."[31]

Since the design of the test relied on interactions between humans and computers, Turing felt the need to include precise details about how they would have entered into communication. To ensure the validity of the test, the interrogator needed to communicate with both human and computer players without receiving any hints about their identities other than the contents of their messages. Communications between humans and computers in the test were thus meant to be anonymous and disembodied.[32] In the absence of video displays and even input devices such as the electronic keyboard, Turing imagined that the answers to the judge's inputs "should be written, or better so, typewritten," the ideal arrangement being "to have a teleprinter communicating between the two rooms."[33] Considering how, as media historians have shown, telegraphic transmission and the typewriter mechanized the written word, making it independent from its author, Turing's solution shows an acute sense of the role of media in communication.[34] The test's technological mediation aspires to make computer and human actors participate in the experiment as pure content, or to use a term familiar to communication theory, as *pure information*.[35]

Before "Computing Machinery and Intelligence," in an unpublished report written in 1947, Turing had fantasized about the idea of creating a different kind of machine, able to imitate humans in their entirety. This Frankenstein-like creature was a cyborg made up of a combination of communication media: "One way of setting about our task of building a 'thinking machine' would be to take a man as a whole and try to replace all the parts of him by machinery. He would include television cameras, microphones, loudspeakers, wheels and 'handling servo-mechanisms' as well as some sort of 'electronic brain.' This would of course be a tremendous undertaking."[36] It is tempting to read this text, notwithstanding its ironic tone, along the lines of Marshall McLuhan's aphorism according to which media are "extensions of man"—in other words, that media provide technological proxies through which human skills, senses, and actions are mechanically reproduced.[37] Although Turing's memo was submitted in 1947, thus almost two decades before the publication of McLuhan's now classic work, Turing's vision was part of a longer imaginative tradition of cyborg-like creatures that presented media such as photography, cinema, and the phonograph as substitutes for human organs and senses.[38] For what concerns the history of AI, Turing's imagination of a machine that takes "man as a whole and tr[ies] to replace all the parts of him by machinery" points to the role of media in establishing the conditions for successful AI.[39] Indeed, the history of the systems developed in the following decades demonstrates the impact of communications media. The use of different technologies and interfaces to send messages informs, if not directly the computational nature of these programs, the outcome of their communication, that is, how they impact on human users. A social media bot on Twitter, for instance, may be programmed to respond with the same script as a bot in a chatroom, but the nature and effects of the communication will be differently contextualized and understood. Similarly, where voice is concerned, the nature of communication will be different if AI voice assistants are applied to domestic environments, like Alexa, embedded in smartphones and other mobile devices, like Siri, or employed to conduct telephonic conversations, as in automated customer services or Google's in-progress Duplex project.[40]

The Turing test, in this sense, is a reminder that interactions between humans and AI systems cannot be understood without considering the circumstances of mediation. Humans, as participants in a communicative interaction, should be regarded as a crucial variable rather than obliterated by approaches that focus on the performance of computing technologies alone. The medium and the interface, moreover, also contribute to the shape of the outcomes and the implications of every interaction. Turing's proposal was, in this sense, perhaps more "Communication Game" than "Imitation Game."

PLAYING WITH TURING

Be it about communication or imitation, the Turing test was first of all a *game*. Turing never referred to it as "test"; this wording took over after his death.[41] The fact that what we now call the Turing test was described by its creator not as a test but as game is sometimes dismissed as irrelevant.[42] Behind such dismissals lies a widespread prejudice that deems playful activities to be distinct from serious scientific inquiry and experiment. In the history of human societies, however, play has frequently been an engine of innovation and change.[43] As concerns the history of AI, games have been a powerful vehicle of innovation, too: AI pioneers such as the father of information theory, Claude Shannon, or the creator of the first chatbot, Joseph Weizenbaum, relied on games to explore and interrogate the meanings and potentials of computing.[44] Turing himself took up chess programming, writing his own chess program on paper in the absence of a computer available to run it.[45]

In 1965, the Russian mathematician Alexander Kronrod was asked to explain why he was using precious computer time at the Soviet Institute of Theoretical and Experimental Physics to engage in a playful, trivial activity such as correspondence chess. He responded that chess was "the drosophila of artificial intelligence."[46] The reference was to *Drosophila melanogaster*, more commonly known as the fruit fly, which was adopted as a "model organism" by researchers in genetics for experimental inquiry in their field.[47] Kronrod suggested that chess provided AI researchers with a relatively simple system whose study could help explore larger questions about the nature of intelligence and machines. The fact that he ultimately lost the directorship of the Institute because of complaints that he was wasting expensive computer resources notwithstanding, his answer was destined to turn into an axiom for the AI discipline. Its pioneers turned their attention to chess because its simple rules made it easy to simulate on a computer, while the complex strategies and tactics involved made it a significant challenge for calculating machines.[48] Yet, as historians of science have shown, such choices are never neutral decisions: for example, the adoption of *Drosophila* as the experimental organism of choice for genetics research meant that certain research agendas became dominant while others were neglected.[49] Similarly, the choice of chess as "AI's drosophila" had wider consequences for the AI field, shaping approaches to programming and particular understandings of what intelligence is. Advances in computer chess privileged the association between intelligence and rational thinking, supported by the fact that chess has been regarded for centuries as evidence of the highest human intelligence.[50]

One wonders if not only chess but games in general are the *Drosophila* of communicative AI. During AI's history, games have allowed scholars and developers to imagine and actively test interactions and communication with machines.[51] While a game can be described in abstract terms as a set of rules describing potential interactions, in pragmatic terms games exist only when enacted—when players make choices and take actions under the constraints of the rules. This applies also to the Imitation Game: it needs players to exist.

Historians of computing have explored the close relationship between games and the emergence of the computer user. In the 1960s, when time-sharing technologies and minicomputers made it possible for multiple individuals to access still scarce computer resources, playing games was one of the first forms of interaction implemented. Games, moreover, were among the first programs coded by early "hackers" experimenting with the new machines.[52] As computer scientist Brenda Laurel puts it, the first programmers of digital games realized that the computer's potential lay not or not only "in its ability to perform calculations but in its capacity to represent action in which humans could participate."[53]

Having computers play games has meaningful consequences for human-computer interaction. It means creating a controlled system where interactions between human users and computer users can take place, identifying the computer as a potential player and, therefore, as an agent within the game's design. This can lead to the distinctions between human and computer players collapsing. Within the regulated system of a game, in fact, a player is defined by the characteristics and the actions that are relevant to the game. It follows that if a computer and a human agent playing the game are compared, they are in principle indistinguishable within the boundaries of that game. This has led game studies scholars to argue that the division between humans and machines in digital games is "completely artificial" since "both the machine and the operator work together in a cybernetic relationship to effect the various actions of the video game."[54]

At different moments in the history of AI, games encouraged researchers and practitioners to envision new pathways of interactions between humans and machines. Chess and other table games, for instance, made possible a confrontation between humans and computers that required turn taking, a fundamental element in human communication and one that has been widely implemented in interface design. Similarly, digital games opened the way for experimenting with more complex interactive systems involving the use of other human senses, such as sight and touch.[55] The Turing test, in this context, envisioned the emergence of what Paul Dourish calls "social computing," a range of systems for human-machine interactions by which understandings of the social worlds are incorporated into interactive computer systems.[56]

At the outset of the computer era, the Turing test made ideas such as "thinking computers" or "machine intelligence" imaginable in terms of a simple game situation in which machines rival human opponents. In his visionary book *God & Golem, Inc.* (1964), the founder of cybernetics, Norbert Wiener, predicted that the fact that learning machines could outwit their creators in games and real-world contests would raise serious practical and moral dilemmas for the development of AI. But the Turing test did not just posit the possibility that a machine could beat a human at a fair play. In addition, in winning the Imitation Game—or passing the Turing test, as one would say today—machines would advance one step further in their impertinence toward their creators. They would trick humans, leading us into believing that machines aren't different from us.

A PLAY OF DECEPTION

In *Monkey Shines*, a 1988 film directed by George Romero, a man named Allan is paralyzed from below the neck by an accident. To cope with the situation, he is offered a trained monkey to help in his daily life. The monkey soon exhibits a startling intelligence and develops an affectionate relationship with the protagonist. Yet, in keeping with Romero's approach to filmmaking—he is well known for the *Night of the Living Dead* saga—the story soon turns into a nightmare. The initial harmony between the animal and its master is broken as the monkey goes on a killing spree, taking out some of Allan's loved ones. Though unable to move, Allan still has an advantage over the monkey: humans know how to lie, a skill that animals—the film tells us—are lacking. Allan succeeds in stopping the monkey by pretending to seek its affection, thus attracting it into a deadly trap.

The idea that the capacity to lie is a defining characteristic of humans resonates in the Turing test. If the test is an exercise in human-computer interaction, this interaction is shaped by trickery: the computer will "win" the game if it is able to deceive the human interrogator. The Turing test can be regarded, in this sense, as a lie-detection test in which the interrogator cannot trust her conversation partner because much of what a computer will say is false.[57] Critics of the Turing test have sometimes pointed to this aspect as one of its shortcomings, reasoning that the ability to fool people should not be posed as adequate test of intelligence.[58] This, however, can be explained by the fact that the test does not measure intelligence, only the capacity to reproduce it. Evaluating AI in terms of imitation, the test poses the problem of AI from the perspective of the user: it is only by focusing on the observer, after all, that one may establish the efficacy of an illusion.[59] Including lying and deception in the mandate of AI becomes thus equivalent with defining machine intelligence in terms of how it is perceived by human users, rather than in absolute terms. Indeed in the decades after Turing's proposal, as computer programs were developed to try their luck with the test, deception became a common strategy: it became evident to programmers that there were strategies to exploit the fallibility of judges, for instance by employing nonsense or irony to deflect questions that could expose the computer program.[60]

Media historian Bernard Geoghegan has suggested that the Turing test's relationship with deception reflects the heritage of cryptological research in early AI.[61] During World War II, Turing contributed to efforts conducted at Bletchley Park to decrypt the ciphered messages used by the Germans and their allies to communicate in secret. This also involved creating electronic machines, including one of the very first digital computers, to enhance the speed and reach of cryptographic work.[62] Computer scientist Blay Whitby, a close acquaintance of Turing, compared the efforts of British cryptographers to those of the human interrogator in the Turing test. Cryptographers had to deduce the functioning of the cryptographic machines created by the Nazi, such as the notorious Enigma, purely by analyzing the ciphered message; in a similar way, Whitby notes, the judge in the Turing test has to find out the human or mechanical identity of her conversation partner only by assessing the "output" of the conversation.[63]

By comparing the dynamics of the Turing test to warfare, the cryptographic interpretation envisages an internal tension between the machine and the human. In fact, the prospect of machines passing the Turing test has often awakened fears of a robot apocalypse and the loss of primacy for humans.[64] Yet there is also an element of playfulness in the deception administered by computers in the Turing test. As I have shown, Turing's insistence that it was to be regarded as a game underpinned its pertinence to the world of play. As proposed by cultural historian Johan Huizinga, play is "a free activity standing quite consciously outside 'ordinary' life as being 'not serious,' but at the same time absorbing the player intensely and utterly."[65] Thus in the Turing test, the computer's deceptive stance is innocuous precisely because it is carried within such a playful frame, with the complicity of human players who follow the rules of the game and participate willingly in it.[66]

Turing presented his test as an adaptation of an original "imitation game" in which the interrogator had to determine who was the man and who the woman, placing gender performance rather than machine intelligence at center stage.[67] The test, in this regard, can be seen as part of a longer genealogy of games that, before the emergence of electronic computers, introduced playful deception as part of their design.[68] The Lady's Oracle, for instance, was a popular Victorian pastime for social soirées. The game provided a number of prewritten answers that players selected randomly to answer questions posed by other players. As amusing and surprising replies were created by chance, the game stimulated the players' all-too-human temptation to ascribe meaning to chance.[69] The "spirit boards" that reproduce the dynamics of spiritualist séances for the purpose of entertainment are another example.[70] Marketed explicitly as amusements by a range of board game companies since the late nineteenth century, they turn spirit communication into a popular board game that capitalizes on the players' fascination with the supernatural and their willingness—conscious or unconscious—to make the séance "work."[71] The same dynamics characterize performances of stage magic, where the audience's appetite for the show is often attributable to the pleasures of falling in with the tricks of the prestidigitator, of the discovery that one is liable to deception, and of admiration for the performer who executes the sleights of hand.[72]

What these activities have in common with the Turing test is that they all entertain participants by exploiting the power of suggestion and deception. A negative connotation is usually attributed to deception, yet its integration in playful activities is a reminder that people actively seek situations where they may be deceived, following a desire or need that many people share. Deception in such contexts is domesticated, made integral to an entertaining experience that retains little if anything of the threats that other deceptive practices bring with them.[73] Playful deception, in this regard, posits the Turing test as an apparently innocuous game that helps people to experiment with the sense, which characterizes many forms of interactions between humans and machines, that a degree of deception is harmless and even functional to the fulfillment of a productive interaction. Alexa and Siri are perfect examples of how this works in practice: the use of human voices and names with which these "assistants" can be summoned and the consistency of their behaviors stimulate users to assign a certain personality to them. This, in turn, helps users to introduce these systems more easily into their everyday lives and domestic spaces, making them less threatening and more familiar. Voice assistants function most effectively when a form of playful and willing deception is embedded in the interaction.

To return to the idea underlined in *Monkey Shines*, my reading of the Turing test points to another suggestion: that what characterizes humans is not as much their ability to deceive as their capacity and willingness to fall into deception. Presenting the possibility that humans can be deceived by computers in a reassuring, playful context, the Turing test invites reflection on the implications of creating AI systems that rely on users falling willingly into illusion. Rather than positing deception as an exceptional circumstance, the playfulness of the Imitation Game envisioned a future in which banal deception is offered as an opportunity to develop satisfactory interactions with AI technologies. Studies of social interaction in psychology, after all, have shown that self-deception carries a host of benefits and social advantages.[74] A similar conclusion is evoked and mirrored by contemporary research in interaction design pointing to the advantages of having users and consumers attribute agency and personality to gadgets and robots.[75] These explorations tell us that by cultivating an impression of intelligence and agency in computing systems, developers might be able to improve the users' experience of these technologies.

The attentive reader might have grasped the disturbing consequences of this apparently benign endeavor. McLuhan, one of the most influential theorists of media and communication, used the Greek myth of Narcissus as a parable for our relationship with technology. Narcissus was a beautiful young hunter who, after seeing his own image reflected in a pool, fell in love with himself. Unable to move away from this mesmerizing view, he stared at the reflection until he died. Like Narcissus, people stare at the gadgets of modern technology, falling into a state of narcosis that makes them unable to understand how media are changing them.[76] Identifying playful deception as a paradigm for conceptualizing and constructing AI technologies awakens the question of whether such a sense of narcosis is also implicated in our reactions to AI-powered technologies. Like Narcissus, we regard them as inoffensive, even playful, while they are changing dynamics and understandings of social life in ways that we can only partially control.

THE MEANING OF A TEST

So much has been written about the Turing test that one might think there is nothing more to add. In recent years, many have argued that the

Turing test does not reflect the functioning of modern AI systems. This is true if the test is seen as a comprehensive test bed for the full range of applications and technologies that go under the label "AI," and if one does not acknowledge Turing's own refusal to tackle the question whether "thinking machines" exist or not. Looking at the Turing test from a different perspective, however, one finds that it still provides exceedingly useful interpretative keys to understanding the implications and impact of many contemporary AI systems.

In this chapter, I've used the Turing test as a theoretical lens to unveil three key issues about AI systems. The first is the centrality of the human perspective. A long time before interactive AI systems entered domestic environments and workspaces, researchers such as Turing realized that the extent to which computers could be called "intelligent" would depend on how humans perceived them rather than on some specific characteristic of machines. This was the fruit, in a sense, of a failure: the impossibility of finding an agreement about definitions of the word *intelligence* and of assessing the machine's experience or consciousness without being "inside" it. But this realization was destined to lead toward extraordinary advancements in the AI field. Understanding the fact that AI is a relational phenomenon, something that emerges also and especially within the interaction between humans and machines, stimulated researchers and developers to model human behaviors and states of mind in order to devise more effective interactive AI systems.

The second issue is the role of communication. Imagined by Turing at a time when the available tools to interact with computers were minimal and the very idea of the user had not yet emerged as such, the Turing test helps us, paradoxically, to understand the centrality of communication in contemporary AI systems. In computer science literature, human-computer interaction and AI are usually treated as distinct: one is concerned with the interfaces that enable users to interact with computing technologies, the other with the creation of machines and program completing tasks that are considered intelligent, such as translating a piece of writing into another language or engaging in conversation with human users. Yet the Turing test, as I have shown in this chapter, provides a common point of departure for these two areas. Regardless of whether this was or was not among Turing's initial intentions, the test provides an opportunity to consider AI also in terms of how the communication between humans and computers is embedded in the system. It is a reminder that AI systems are not just computing machines but also media that enable and regulate specific forms of communication between users and computers.[77]

The third issue is related to the central preoccupation of this book: the relationship between AI and deception. The fact that the Turing test posited a situation in which a human interrogator was prone to deception by the computer shows that the problem of deception sparked reflections in the AI field already in its gestational years. Yet the game situation in which the test is framed stimulates one to consider the playful and the apparently inoffensive nature of this deception. As discussed in the introduction, after all, media technologies and practices, including stage magic, trompe l'oeil painting, cinema, and sound recording, among many others, are effective also to the extent in which they open opportunities for playful and willing engagement with the effects of deception.[78] The playful deception of the Turing test, in this sense, further corroborates my claim that AI should be placed within the longer trajectory of deceitful media that incorporate banal deception into their functioning.

CHAPTER 2

How to Dispel Magic

Computers, Interfaces, and the Problem of the Observer

In the summer of 1956 a group of mathematicians, computer engineers, physicists, and psychologists met for a brainstorming workshop in Dartmouth, US to discuss a new research area they ambitiously called "artificial intelligence." Although no particular breakthrough was reached at the conference, it was there that some of the scientists who shaped the new field in the years ahead—such as Marvin Minsky, John McCarthy, Allen Newell, Herbert Simon, and Claude Shannon—defined the objectives that would lead AI efforts.[1] All shared the persuasion that human intelligence was based on formalized reasoning whose dynamics could be reproduced successfully by digital computers. If asked the question that Turing had cautiously dismissed few years earlier—"Can machines think?" —most of them would have probably answered: Yes, and we will soon have enough evidence to persuade even the most skeptical inquirers.

More than sixty years later, evidence supporting this view is still uncertain and ambivalent. Despite their excessive optimism, however, Dartmouth's AI pioneers established a foundation for exceptional development in computer science that continues to exercise an influence today. Between the end of the 1950s and the early 1970s, they facilitated a range of practical achievements that dramatically raised expectations about the prospects of the field. It was a period of exceptional enthusiasm, described by historians as the "golden age" of AI.[2] Thriving technologies such as language processing, machine translation, automated problem-solving, chatbots, and computer games all originated in those years.

Deceitful Media. Simone Natale, Oxford University Press (2021). © Oxford University Press.
DOI: 10.1093/oso/9780190080365.003.0003

In this chapter, I look at this early AI research to unveil a less evident legacy of this pioneering period. I tell the story of how members of the AI community realized that observers and users could be deceived into attributing intelligence to computers, and how this problem became closely related to the building of human-computer interactive systems that integrated deception and illusion in their mechanisms. At first glance, this story seems in contradiction with the fact that many of the researchers who shaped this new field believed in the possibility of creating "strong" AI, that is, a hypothetical machine that would have the capacity to complete or learn any intellectual task a human being could. Yet practical explorations in AI also coincided with the discovery that humans, as Turing anticipated, are part of the equation that defines both the meaning and the functioning of AI.

In order to tell this story, one needs to acknowledge that the development of AI cannot be separated from the development of human-computer interactive systems. I look, therefore, at the history of AI and human-computer interaction in parallel. I start by examining the ways AI researchers acknowledged and reflected on the fact that humans assign intelligence to computers based on humans' biases, visions, and knowledge. As scientists developed and presented to the public their computing systems under the AI label, they noted that observers were inclined to exaggerate the "intelligence" of these systems. Because early computers offered few possibilities of interaction, this was an observation that concerned mostly the ways AI technologies were presented and perceived by the public. The development of technologies that created new forms and modalities of interaction between humans and computers was, in this regard, a turning point. I move on therefore to discuss the ways early interactive systems were envisioned and implemented. Drawing on literature in computer science and on theories in media studies that describe computer interfaces as illusory devices, I show that these systems realized in practice what the Turing test anticipated in theory: that AI only exists to the extent it is perceived as such by human users. When advancements in hardware and software made new forms of communication between humans and computers possible, these forms shaped practical and theoretical directions in the AI field and made computing a key chapter in the story of the lineage of deceitful media.

BUILDING TECHNOLOGY, CREATING MYTHS

Watching moving images on the screens of our laptops and mobile devices is today a trivial everyday experience. Yet one can imagine the marveling of spectators at the end of the nineteenth century when they saw for the first time images taken from reality animated on the cinematic screen. To many early observers, film appeared as a realm of shadows, a feat of magic similar to a conjurer's trick or even a spiritualist séance. According to film historians, one of the reasons that early cinema was such a magical experience was the fact that the projector was hidden from the spectators' view. Not seeing the source of the illusion, audiences were stimulated to let their imaginations do the work.[3]

At the roots of technology's association with magic lies, in fact, its opacity. Our wonder at technological innovations often derives from our failure to understand the technical means through which they work, just as our amazement at a magician's feat depends in part on our inability to understand the trick.[4] From this point of view, computers are one of the most enchanting technologies humans have ever created. The opacity of digital media, in fact, cannot be reduced to the technical skills and knowledge of users: it is embedded in the functioning of computing technologies. Programming languages feature commands that are intelligible to computer scientists, allowing them to write code that executes complex functions. Such commands, however, correspond to actual operations of the machine only after having been translated multiple times, into lower-level programming languages and finally into machine code, which is the set of instructions in binary numbers executed by the computer. Machine code is such a low level of abstraction that it is mostly incomprehensible to the programmer. Thus, even developers are unable to grasp the stratifications of software and code that correspond to the actual functioning of the machine. Computers are in this sense the ultimate black box: a technology whose internal functioning is opaque even to the most expert users.[5]

It is therefore unsurprising that AI and more generally electronic computers stimulated from their inception a plethora of fantasies and myths. The difficulty of figuring out how computers worked, combined with overstated reports about their achievements, sparked the emergence of a vibrant imaginary surrounding the new machines. As historian of computing C. Dianne Martin demonstrated, based on a body of poll-based sociological evidence and content analysis of newspapers, a large segment of public opinion in the 1950s and early 1960s came to see computers as "intelligent brains, smarter than people, unlimited, fast, mysterious, and frightening."[6] The emergence of AI was deeply intertwined with the rise of a technological myth centered around the possibility of creating thinking machines. This did not go unnoticed by the scientists who built the AI field in the decades after World War II. As early achievements were given coverage in the press and discussed in public forums, researchers found out that if a computer system was presented as intelligent, people had a strong tendency to perceive it as such. Despite the fact that the systems the pioneers of the discipline developed were still of limited complexity and practical use, they seemed to many to be powerful demonstrations that scientists were advancing computers toward being capable of exercising humanlike faculties such as intuition, perception, and even emotion.

According to Martin, mainstream journalists shaped the public imagination of early computers by means of misleading metaphors and technical exaggerations. By contrast, computer scientists attempted to "dispel developing myths about the new devices."[7] Yet the response of many AI researchers was more ambivalent than Martin, herself a computer scientist, concedes. As researchers were adopting a new terminology to describe functions and methods relevant to AI systems, they often selected metaphors and concepts that suggested a close resemblance to human intelligence.[8] One of the books that mapped the territory, for instance, cautiously defined AI in 1968 as "the science of making machines do things that would require intelligence if done by men," implicitly reinstating the distinction between the mechanic and the human.[9] Yet contributors to this book described computing processes in terms of concepts closely associated with human intelligence.[10] Words such as "thinking," "memory," and even "feeling" masked the fact that programmers were most often aiming to create systems that only *appeared* intelligent. The use of analogies is widespread in scientific discourse, and it became particularly prominent in the transdisciplinary research approach adopted by AI researchers, in which computer science intersected with fields such as psychology, biology, and mathematics, among others.[11]

Although early AI systems had generated promising outcomes, the differences from how humans think were evident. For instance, elements such as intuition or fringe consciousness, which play important roles in human intelligence, were completely absent in the computers' formal models.[12] That misleading metaphors were nonetheless used depended only in part on the tendency for overstatement that characterized the communication of much early AI research: it was also due to the theoretical assumptions that underpinned AI. The underlying theory in the initial phase of AI was that rational thinking was a form of computation, and that mental processes could not just be described symbolically—they were symbolic.[13] This hypothesis came together with the belief that computers could replicate the workings of the human mind.

One of the things that made this hypothesis convincing was the fact that not only computer technologies but also everything that has to do with intelligence is by definition opaque. If it is difficult to bridge the gap between what software achieves and what happens at the level of computer circuits, it is even more difficult to imagine how our thoughts and emotions correspond to specific physical states inside the human brain and body. We know, in fact, very little about how the brain works.[14] It was in the space between the two sides of this double lack of understanding that the possibility opened up for imagining computers as *thinking machines* even when little evidence was available to suggest a close similarity between computers and human brains. The opacity of both computers and human brains meant that the statement "computers are thinking machines" was impossible to prove but also impossible to dismiss. Consequently, the dream of building an electronic brain drove the efforts of many AI researchers during a phase of extraordinary expansion of computing technologies.

Some researchers saw the excessive enthusiasm around early AI systems as a form of deception. Cognitive scientist Douglas Hofstadter reflected on such dynamics by pointing out that "to me, as a fledgling [AI] person, it was self-evident that I did not want to get involved in that trickery. It was obvious: I don't want to be involved in passing off some fancy program's behavior for intelligence when I know that it has nothing to do with intelligence."[15] Other AI researchers were more ambivalent. Most scientists usually expressed caution in academic publications, but in platforms aimed at nonacademic audiences and in interviews with the press they were far less conservative.[16] Marvin Minsky at the Massachusetts Institute of Technology (MIT), for instance, suggested in a publication aimed at the broad public that once programs with capacity for self-improvement were created, a rapid evolutionary process would lead to observing "all the phenomena associated with the terms 'consciousness,' 'intuition' and 'intelligence' itself."[17]

No one in the field could ignore how important it was to navigate the of public expectations, perceptions, and fantasies that surrounded AI.[18] The tendency to overstate actual and potential achievements of AI was fueled by the need to promise exciting outcomes and practical applications that would attract funding and attention to research. Reporting on his experience as an AI pioneer in the 1950s, Arthur L. Samuel recalled: "we would build a very small computer, and try to do something spectacular with it that would attract attention so that we would get more money."[19] The myth of the thinking machine also functioned as a unifying goal for researchers in the nascent AI field. As historians of technology have shown, future-oriented discourse in techno-scientific environments contributes to a shift of emphasis away from the present state of research and toward an imagined prospect when the technology will be successfully implemented.

This shift facilitates the creation of a community of researchers, introducing a shared objective or endpoint that informs and organizes the work of the scientists, technologists, and engineers involved in such a community.[20]

As practical systems were implemented with mixed results, it became clear that the race for AI was being run not only in the realm of technology but also in the realm of the imagination. The idea of thinking machines that would rival human intelligence kindled the fascination of the public, stimulating the emergence of dreams as well as nightmares about the future of AI. These sentiments reverberate today in visions about the singularity or the upcoming robot apocalypse.[21] When science fiction writer Arthur C. Clarke, who developed in collaboration with Stanley Kubrick the subject for *2001: A Space Odyssey* and wrote the novel with the same title, asked his friend Marvin Minsky to consult on how to imagine an intelligent computer on board a spaceship, the MIT computer scientist was ready to assist. As his team at MIT had produced a number of systems that despite their limited functionality could be interpreted by some to be real "intelligence," nobody was a better fit to inspire the technological imagination of Kubrick's audience.[22]

BRINGING COMPUTERS DOWN TO EARTH

The emergence of AI as a scientific field thus coincided with the discovery that popular myths about computers affect how AI technologies are perceived and interpreted by the public. Yet, as long as computer access was limited to a small community of experts, as was the case until at least the 1950s, the problem of how people perceived AI systems was more speculative than practical. Although "computer brains" and humanoid robots were capturing the public imagination, few people were able to observe directly how AI worked, let alone interact with AI technologies. Things, however, were moving fast. In the late 1950s and during the 1960s, a series of theoretical and practical innovations opened new pathways toward interaction with computers. These efforts brought with them a renewed emphasis on the role of human users in computing systems, which would in turn inform reflections and practical applications relating to the role of the human component in AI.

One of the most popular ideas in McLuhan's media theory is that media are "extensions of man" (and of woman, we should add).[23] The most common interpretation of this concept is that new media change humans at an anthropological level, affecting how individuals access the world as well as the scale or pattern of human societies. McLuhan argues, for instance, that the wheel serves as an extension of human feet, accelerating movement and exchange across space. As a result, the wheel has enabled new forms of political organizations, such as empires, and has changed humans at a social, cultural, and even anthropological level. Seen from a complementary perspective, however, the notion that media are extensions of humans also suggests another important aspect of media: that they are meant to fit humans. Media are envisioned, developed, and fabricated so that they can adapt to their users—in McLuhan's words, to become their extensions. Paraphrasing an old text, *so humankind created media in its image, in the image of humankind it created them . . .*

It is easy to see how this applies to different kinds of media. The invention of cinema, for instance, was the result not only of prodigious engineering efforts but also of decade-long studies about the functioning of human perception. Knowledge about vision, perception of movement, and attention was incorporated into the design of cinema so that the new medium could provide an effective illusion and entertain audiences around the world.[24] Likewise, sound media, from the phonograph to the MP3, have been constructed in accordance with models of human hearing. In order to improve capacity while retaining quality of sound, frequencies that are outside the reach of human hearing have been disregarded, adapting technical reproduction to what and how we actually hear.[25] The problem was not so much how an early phonograph cylinder, a vinyl record, or an MP3 sounded in physical terms; it was how they sounded *for humans*. As proposed in this book, this is why all modern media incorporate banal deception: they exploit the limits and characteristics of human sensoria and psychology in order to create the particular effects that suit their intended and prepared use.

Computers are in this regard no exception. Throughout the history of computing, a plethora of different systems have been developed that have enabled new forms of communication with users.[26] In this context, the computer has become more and more evidently an extension of humans, hence a communication medium. The need for more accessible and functional pathways of interaction between computers and human users stimulated computer scientists to interrogate the ways humans access and process information, and to apply the knowledge they gathered in the development of interactive systems and interfaces. Not unlike what happened with cinema or sound media, this effort entailed exploring ways computers could be made "in the image of humankind."

While the history of AI is usually written as distinct from the history of human-computer interaction, one need only look at AI's early development to realize that no such rigid distinction is possible.[27] The golden age of AI coincided not just chronologically with the emergence of interactive computer systems as a specific area of inquiry in computer sciences. In laboratories and research centers across the United States, Europe, Russia, and Japan, the goal of creating "intelligent" machines came together with the objective of implementing interactive systems ensuring wider, easier, and more functional access to computers. This was openly acknowledged by computer scientists of the time, who considered human-machine systems largely within the remits of AI.[28] The division between AI and human-computer interaction is thus the fruit of a retrospective partition, rather than a useful organizing principle to understand the history of computer science.[29]

The overall vision that shaped the development of early interactive systems was "human-computer symbiosis," a concept originally proposed by J. R. C. Licklider in an article published in 1960 that achieved the status of a unifying reference point for computer and AI research.[30] The article used the concept of symbiosis—a metaphor taken from biology—to propose that real-time interactions would pave the way for a partnership between humans and computers. Licklider believed that this partnership, if implemented through practical developments in computer hardware and software, would "perform intellectual operations much more effectively than man alone can perform them."[31] Katherine Hayles has influentially argued that the emergence of human-computer symbiosis in postwar computing research broke down the distinction between humans and machines, giving away the notion of a "natural" self and reconceptualizing human intelligence as coproduced with intelligent machines.[32] Licklider's vision of symbiosis between humans and computers, however, points in another direction as well. Similar in this regard to the Turing test, this vision entailed acknowledging that computing machines were to be understood not only on their own but also through their interactions with human users.

Although the term *interaction* was usually privileged, the new paradigm of symbiosis rested on the assumption that humans and computers could enter into communication with each other. The problem was that communication was a heated concept in AI and computer science. Information theory and cybernetics privileged a disembodied understanding of communication that placed little or no emphasis on the meaning and context of communications.[33] This was integral to a model that foregrounded the equivalence between brains and computers, each seen as an information-processing device whose working could in principle be described in abstract, mathematical terms.[34] Understanding communication as disembodied provided a powerful symbolic tool to envision, project, and implement human-computer systems in terms of a feedback mechanism between two entities that spoke, so to speak, the same language. At the same time, however, this approach conflicted with practical experiences with these systems. When humans are involved, in fact, communication is always situated within a sociocultural milieu. As Lucy Suchman would later make clear, any effort to create human-computer interactions entails inserting computer system within real-world situations.[35]

In the space between these two understandings of communication—disembodied and abstract on the one side, as in the theoretical foundations of computer science, embodied and immersed in social situations on the other side, as in practical experiences with new interactive systems—one can better understand the apparent contradiction that characterized the exploration of many AI pioneers. Leading scientists embraced the myth of the "thinking machine," which was rooted in the idea that the brain is a machine and the activity of the neurons can be described mathematically, just as the operations of computers. At the same time, the experience of creating AI applications and putting them into use led scientists to consider the specific perspectives of human users. In order to achieve Licklider's goal of a symbiotic partnership performing intellectual operations more effectively than humans alone, the complex processes of electronic computing were to be adapted to users. Computers, in other words, were to be brought "down to earth," so to speak, so that they could be easily accessed and used by humans.

As systems were developed and implemented, researchers realized that human psychology and perception were important variables in the development of effective human-computer interactions. Early studies focused on issues such as response time, attention, and memory.[36] The development in the 1960s of time-sharing, which allowed users to access computer resources in what they perceived as real time, is a case in point.[37] One of the leading AI scientists, Herbert Simon, contended in his 1966 article "Reflections on Time Sharing from a User's Point of View" that a time-sharing system should be considered "fundamentally and essentially a man-machine system whose performance depends on how effectively it employs the human nerveware as well as the computer hardware."[38] Consequently, time-sharing had to be modeled against the ways humans receive and process information. Likewise, for Martin Greenberger, who researched both AI and time-sharing at MIT, computers and users were to be considered "the two sides of time sharing." Only a perspective that looked at both sides would lead to the development of functional systems.[39] In this regard, he observed, time-sharing was "definitely a concession to the user and a recognition of his point of view."[40]

It is important to remember that the "user" to which researchers such as Greenberger were referring was not any kind of user. Until at least the emergence of personal computing in the late 1970s and early 1980s, human users were mainly the mathematicians and engineers who were designing and running the earliest computer systems. With the dream of human-computer symbiosis gaining ground, conceptualizations of users started to expand to the "executive" but still maintained a very narrow focus on white, educated men. This corresponded in AI to representing the "human," to whose image intelligent systems were to be calibrated, in exceedingly restrictive terms concerning race, gender, and class—a mystification that has resulted, throughout the history of computer science, in forms of bias and inequality that have informed computing technologies up to the present day.[41]

In the climate of burgeoning AI research, the new emphasis on the perspective of the user stimulated researchers to ask why and in which conditions people consider machines "intelligent." Artificial intelligence researchers began to understand that since communications between humans and machines were situated in a social and cultural context, attributions of intelligence to computers could also emerge as the result of illusion and deception. The diffusion of myths about thinking machines and computer brains, in this context, increased the possibility of relatively simple computing systems being perceived as intelligent by many, exaggerating the achievements of the field.

As technologies such as time-sharing and other human-computer interaction systems were implemented with increasingly successful outcomes, this problem started to preoccupy leading AI scientists such as Minsky. As reported by his pupil Joseph Weizenbaum, "Minsky has suggested in a number of talks that an activity which produces results in a way which does not appear understandable to a particular observer will appear to that observer to be somehow intelligent, or at least intelligently motivated. When that observer finally begins to understand what has been going on, he often has a feeling of having been fooled a little. He then pronounces the heretofore 'intelligent' behavior he has been observing as being 'merely mechanical' or 'algorithmic.' "[42] In the proceedings of a 1958 symposium on the "Mechanization of Thought Processes," Minsky admitted that machines may seem more resourceful and effective than they actually are. He further elaborated that this depends not only on the machine's functioning but also on "the resources of the individual who is making the observation."[43] In theory, Minsky believed in the principle that the behavior of a machine is always explicable in mechanistic terms, as a consequence of its past states, internal structure, external contingencies, and the relations between them. In practice, however, he conceded that a machine invites different interpretations from different observers. A person with little insight into computing and mathematics, for example, may see intelligence where a more expert user wouldn't. A similar dynamic, after all, also characterizes judgments about intelligence and skill in humans, which "are often related to our own analytic inadequacies, and . . . shift with changes in understanding."[44] In another article, Minsky therefore argued that "intelligence can be measured only with respect to the ignorance or lack of understanding of the observer."[45]

These were not mere footnotes based on casual observations. Acknowledging the contribution of the observer—described as an outsider who approaches computers with limited knowledge of AI—signaled the realization that AI only existed within a cultural and social environment in which the user's perceptions, psychology, and knowledge played a role. The fact that this was recognized and discussed by Minsky—a leading AI scientist who firmly believed that the creation of thinking computers was a definite outcome to be achieved in a relatively close future—makes it all the more significant.[46] Although acknowledging that observers wrongly attribute intelligence to machines might seem at odds with Minsky's optimism for AI's prospects, the contradiction evaporates if one considers that AI emerged in close relationship with the development of human-computer interactive systems. In a context wherein human users are increasingly exposed to and interacting with computers, believing in the possibility of creating thinking machines is entirely compatible with acknowledging that intelligence can be the fruit of an illusion. Even while professing full commitment to the dream of creating thinking machines, as Minsky did, one could not dismiss the perspective of humans as part of the equation.

In the 1960s, an age where both AI and human-computer interaction posed firm foundations for their subsequent development, the problem of the observer—that is, the question of how humans respond to witnessing machines that exhibit intelligence—became the subject of substantial reflection. British cybernetician and psychologist Gordon Pask, for instance, pointed out in a 1964 computer science article that "the property of intelligence entails the relation between an observer and an artifact. It exists insofar as the observer *believes* that the artifact is, in certain essential respects, like another observer."[47] Pask also noted that an appearance of self-organization entails some forms of ignorance on the part of the observer, and explained at length situations in which computing systems could be programmed and presented so that they looked "amusingly lifelike."[48] Besides theoretical debates, actual artifacts were developed for experiments with observers' reactions to machines that displayed some forms of intelligence. At Bell Labs, for instance, the father of information theory, Claude Shannon, and the engineer David Hagelbarger built two computing machines that played simple games through look-ahead principles that predicted human opponents' choices based on the humans' previous moves. The machines exhibited an impressive series of victories and looked intelligent to many, despite their simple design, which exploited the fact that humans are predictable because of their poor ability to generate random patterns. Calling their creations SEER (for "SEquence Extracting Robot") and Mind Reading Machine, Hagelbarger and Shannon half-jokingly attributed a veil of omniscience to them. They had grasped the importance of playing with people's imaginations in order to achieve "AI."[49]

The emphasis on the observer corresponded to the new focus that research in other areas, especially quantum physics, was giving to the role of observation in informing the outcome of experiments. Discoveries in quantum mechanics by scientists such as Werner Heisenberg, Niels Bohr, and Erwin Schrödinger had revealed that the mere observation of a phenomenon inevitably changes that phenomenon. This shattered the Newtonian certainty that physical phenomena were external and independent from observers. Such discoveries, although they only applied to the level of quantum mechanics, which describes nature at the smallest scale of atoms and subatomic particles, sparked reflections about the role of observers in scientific research in other fields, from the social sciences to the natural sciences, and informed cybernetics and AI, as a key publication in this area recognized in 1962.[50]

In summary, the move, sparked by human-computer interaction, to adapt computers to human users corresponded, in the field of AI, to the realization that the supposed intelligence of computers might also be the fruit of the failure of users and observers to understand how computers work. Although the overarching climate of optimism and enthusiasm in the field had marginalized the approach favored by Turing, some of the questions the Turing test had posed were thus integrated within the dominant AI paradigm by the end of the 1960s.

Despite concerns about mystifications of the actual achievements and functioning of computing, researchers in the AI field would embrace the emergent tendency to normalize deception and make it functional to the implementation of any human-computer interaction. In due time, this tendency would pave the way for the rise of the rhetoric of user-friendliness that has dominated the computer industry since the 1980s.

THE MEANINGS OF TRANSPARENCY, OR HOW (NOT) TO DISPEL MAGIC

Early efforts to convey the significance of AI to the public had shown the resilience of popular myths that equated computers with "electronic brains." Rather than being understood as rational machines whose amazing promise remained in the sphere of science, computers were often represented as quasi-magical devices ready to surpass the ability of humans in the most diverse areas.

In this context, the nascent AI community saw the emergence of human-computer interaction systems as a potential opportunity to influence the ways computers were perceived and represented in the public sphere. In an issue of *Scientific American* dedicated to AI in 1966, AI pioneer John McCarthy proposed that the new interactive systems would provide not only greater control of computers but also greater understanding among a wide community of users. In the future, he contended, "the ability to write a computer program will become as widespread as the ability to drive a car," and "not knowing how to program will be like living in a house full of servants and not speaking their language." Crucially, according to McCarthy, the increased knowledge would help people acquire more control over computing environments.[51]

McCarthy's vision posed a potential solution to the problem of the observer. That some users tended to attribute unwarranted intelligence to machines, in fact, was believed to depend strictly on lack of knowledge about computing.[52] Many members of the AI community were therefore confident that the deceptive character of AI would be dispelled, like a magic trick, by providing users with a better understanding of the functioning of computer systems. Once programming was as widespread as driving a car, they reasoned, the problem would evaporate. Artificial intelligence scientists such as McCarthy imagined that interactive systems would have helped achieve this goal, making computers more accessible and better understood than ever before.

Such hope, however, did not take into account that deception, as the history of media shows, is not a transitional but rather a structural component of people's interactions with technology and media. To create aesthetic and emotional effects, media need users to fall into forms of illusion: fiction, for instance, stimulates audiences to temporarily suspend their disbelief, and television provides a strong illusion of presence and liveness.[53] Similarly, the interaction between humans and computers is based on interfaces that provide a layer of illusion concealing the technological system to which they give access. The development of interactive computing systems meant therefore that magic and deception, rather than being dismissed, were incorporated into interface design.[54]

Besides symbiosis, computer scientists employed a plethora of other metaphors to describe the effects of early interactive systems.[55] In his article on time-sharing cited earlier, Martin Greenberger chose a particularly interesting image. In order to corroborate his point that the perspectives of both computers and users needed be taken into account, he compared time-sharing to vision glasses.[56] This optical metaphor underlined that because both sides of the interaction "saw" the process through a different perspective, engineers and designers needed to calibrate the lenses of both components to help them navigate a common environment and enter into interaction with each other. Optical media, however, have been studied and developed since antiquity to perform illusions and tricks. In this sense, Greenberger's comparison may suggest that human-computer interaction had something in common with the way a conjurer manipulates the perceptions of audiences to create a particular effect on the stage. Time-sharing is, after all, a form of illusion itself, which conceals the internal functioning of computers and adapts it to human users' perceptions of time and synchronicity.[57] The ground for the development of time-sharing is that computers are faster than humans. Although a computer is actually handling one process at a time, taking turns between them, thanks to time-sharing users have the impression of continuity of their own interaction. As McCorduck points out, it is "really a trick based on the mismatch between the slowness of human reaction and the speed of the computer."[58]

Scholars in media studies have suggested that not only time-sharing but all interfaces essentially rely on deception. In technical terms, an interface is a point of interaction between any combination of hardware and software components, but the word is generally used to describe devices or programs that enable users to interact with computers.[59] Because an interface produces a mediation between different layers, Lori Emerson suggests, the interface grants access but "also inevitably acts as a kind of magician's cape, continually revealing (mediatic layers, bits of information, etc.) through concealing and concealing as it reveals."[60] Similarly, Wendy Hui Kyong Chun observes that an interface builds a paradox between invisibility and visibility. Graphic user interfaces, for instance, "offer us an imaginary relationship to our hardware: they do not represent transistors but rather desktops and recycling bins."[61] Fabricating computer interfaces, in this sense, entails creating imaginary worlds that hide the underlying structure of the technical systems to which they give access.[62]

The interface's work of granting access to computers while at the same time hiding the complexity of the systems relates to the way, discussed in the previous chapter, AI facilitates a form of playful deception. Through the game dynamics envisioned by Turing, the illusion of intelligence was domesticated to facilitate a productive interaction between human and machine. Similarly, one of the characteristics of computer interfaces is that they are designed to bring into effect their own illusory disappearance, so that users do not perceive friction between the interface and the underlying system.[63] The illusion, in this context, is normalized to make it appear natural and seamless to users. It was for this reason that the implementation of human-computer interaction coincided with the discovery that where humans are concerned, even things such as misunderstanding and deception inform the new symbiosis.

In the 1966 *Scientific American* issue on AI, Anthony G. Oettinger brought together McCarthy's dream of increased access to computers with his own, different views about software interfaces. Stressing the necessity for easier systems that would make computing comprehensible to everybody, he proposed the concept of "transparent computers" to describe what he considered to be one of the key goals of contemporary software engineering: turning the complex processes of computing hardware into "a powerful instrument as easy to use as pen and paper."[64] Once interest in what happened inside the machine gave place to interest in what could be achieved with computers, he predicted, computers would be ready to serve an enormous variety of functions and roles. Quite remarkably, in the same article he described the computer as a "very versatile and convenient *black box*."[65] This concept, as noted, describes technological artifacts that provide little or no information about their internal functioning. Reading this in its literal sense, one should assume that computers were to be made transparent and opaque at the same time. Of course, Oettinger was playing with a secondary connotation of the word "transparent," that is, easy to use or to understand. That this transparency was to be achieved by making the internal functioning of computers opaque is one of the ironic paradoxes of modern computing in the age of user-friendliness.

From the 1980s, the computer industry, including companies such as IBM, Microsoft, and Apple, embraced the idea of transparent design. Faithful in this regard to McCarthy's call of two decades earlier, they aimed to make computers as widespread as cars; they diverged, however, on the means to achieve this goal. Rather than improving computer literacy by teaching everybody to program, they wanted to make computers simpler and more intuitive so that people with no expertise could use (and buy) them. "Transparent" design did not refer to software that exposed its inner mechanisms but rather software interfaces that concealed the computers' functioning to provide users with an easy, intuitive model.[66] The purpose of transparent design was to make computer interfaces invisible and seamless and thus, according to its supporters, more natural and integrated to users' physical lives. Yet, as in a feat of magic, this peculiar connotation of transparency passes through the interplay between the technical complexity inside the machine and showing a highly controlled environment on the surface.[67] Transparent design sublimates computer systems with a series of representations, embedded in the interface, that help users understand the device's operations only to the degree needed in order to use it. As Lori Emerson notes, the contemporary computing industry sells a dream in which the boundary between human and machine is eradicated through "sophisticated sleights of hand that take place at the level of the interface."[68] Making computers user-friendly is, in this sense, an exercise of "friendly" deception on users. It is a trick to which we voluntarily surrender when we use any of our digital gadgets, from laptops to tablets, from e-readers to smartphones.

Apple, IBM, and other giants of the computer industry were not first in taking up this approach. Weird and contradictory as this concept might be, "opaque transparency" started to shape the development of software interfaces in the 1960s and continues today. Transparency became the organizing principle for producing interfaces that enable seamless interaction with users.[69] Meanwhile in the burgeoning AI community, the dream of dispelling the magic aura of computers was to be superseded by the realization that users' perceptions of computing systems could be manipulated in order to improve interactions between humans and machines. Interactive AI systems, in fact, are themselves interfaces that construct layers of illusion and conceal the functioning of the machine. The users' tendency to attribute intelligence to machines (which had been recognized and discovered by AI researchers such as Minsky) can thus be domesticated and exploited through specific mechanisms of interaction design. Making computers "transparent," as Oettinger proposed, has been achieved within the framework of banal deception.

Computing machines have been from the very beginning part of a lively culture of wonder: from nineteenth-century automata to the emergence of electronic computers, a culture of exhibition and representation has provided constant nourishment to the myths of thinking machines and the association between brains and computers.[70] Yet the feeling of magic that computers inspire has to do not only with the exaggeration of newspaper reports or the popular dreams of science fiction. This feeling is engrained in both sides of the human-computer interaction. On the one side, this feeling of magic is rooted in people's relationships with objects and people's natural tendency to attribute intelligence and sociality to things. On the other side, this feeling is embedded in the functioning of computer interfaces that emerge through the construction of layers of illusion that conceal (while making "transparent") the underlying complexity of computing systems. For this reason, even today, computer scientists remain aware of the difficulty of dispelling false beliefs when it comes to perceptions and uses of AI.[71] The magic of AI is impossible to dispel because it coincides with the underlying logic of human-computer interaction systems. The key to the resilience of the AI myths, then, is to be found not in the flamboyant myths about mechanical brains and humanoid robots but in the innocent play of concealing and displaying that makes users ready to embrace the friendly, banal deception of "thinking" machines.

Developers in computer science would introduce the possibility of deception within a wider framework promising universal access and ease of use for computing technologies, which informed work aimed at improving human-computer interaction. Their explorations entailed a crucial shift away from considering deception something that can be dispelled by improving knowledge about computers and toward the full integration of forms of banal deception into the experience of users interacting with computers. As the next chapter shows, the necessity for this shift became evident in the earliest attempts to create AI systems aimed at interaction with the new communities of users accessing computers through time-sharing systems. One of them, the chatbot ELIZA, not only stimulated explicit reflections on the relationship between computing and deception but also became a veritable archetype of the deceitful character of AI. If the Turing test was a thought experiment contrived by a creative mind, ELIZA was the Turing test made into an actual artifact: a piece of software that became a cause célèbre in the history of AI.

CHAPTER 3
The ELIZA Effect
Joseph Weizenbaum and the Emergence of Chatbots

If asked about ELIZA, the first chatbot ever created, Apple's AI assistant Siri—or at least, the version of Siri installed in my phone—has an answer (fig. 3.1).

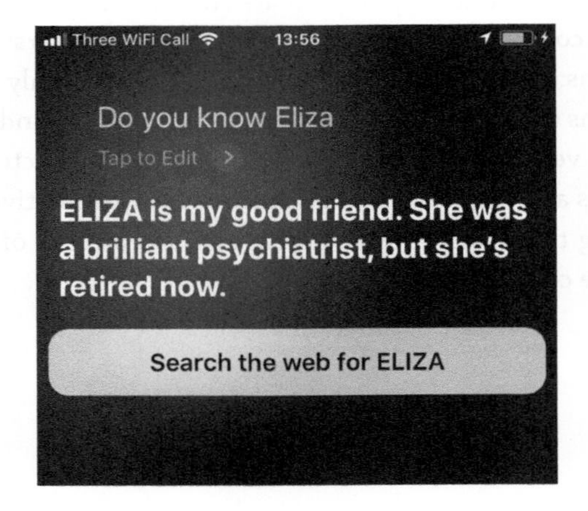

Figure 3.1 Author's conversation with Siri, 15 March 2018.

The answer frames Siri within a long historical trajectory of experiences and experiments with AI programs designed to conduct conversations with humans.[1] ELIZA is in fact widely considered the first chatbot of history, or as Andrew Leonard put it, "bot erectus—the first software

Deceitful Media. Simone Natale, Oxford University Press (2021). © Oxford University Press.
DOI: 10.1093/oso/9780190080365.003.0004

program to impersonate a human being successfully."[2] Its capacity to pass as human, at least in some situations, made it a source of inspiration for generations of developers in natural language processing and other AI applications that engage with language, up to contemporary systems such as Siri and Alexa.

This chapter examines the creation of ELIZA at MIT in 1964–1966 and interrogates its legacy by examining the lively debate it stimulated in the AI community and the public sphere. Often discussed in passing in historical scholarship on AI, computing, and media, the case of ELIZA deserves much more dedicated attention. In fact, the creation and reception of Weizenbaum's ELIZA was a crucial moment for the history of digital media, not only because the program is widely regarded as the first to conduct a conversation in natural language but also because ELIZA was a disputed object that became the center of competing narratives shaping key controversies and discourses about the impact of computing and digital media.

Focusing on the discussions about how to interpret the functioning and success of ELIZA, I argue that it was influential not so much at a technical but mainly at a discursive level. Providing evidence of how relatively simple AI systems could deceive users and create an impression of humanity and intelligence, ELIZA opened up new perspectives for researchers interested in AI and in conversational agents such as chatbots. While the program was relatively unsophisticated, especially compared to contemporary systems, the narratives and anecdotes circulating about it prompted reflection on how humans' vulnerability to deception could be exploited to create efficient interactions between humans and machines. As such, ELIZA is an emblematic example of how artifacts, and in particular software, have social and cultural lives that concern their material circulations as much as the narratives that emerge and circulate about them.

With the creation of ELIZA, the German-born computer scientist Joseph Weizenbaum was determined to stress the illusory character of computers' intelligence. Yet some of the narratives emerging from the creation of ELIZA reinforced the idea that machines and humans think and understand language in similar ways. Consequently, the program was interpreted as evidence in favor of two different, even contrasting visions: on the one side that AI provided only the appearance of intelligence; on the other that AI might actually replicate intelligence and understanding by artificial means. In the following decades, moreover, the mechanisms of projection that inform the use of chatbots and other AI agents came to be described as the "ELIZA effect." In this sense, ELIZA is more than a piece of obsolete software: it is a philosophical toy that, like the optical devices invented in the nineteenth century to disseminate knowledge about illusion and perception, is still a reminder that the power of AI originates in the technology as much as in the perception of the user.

THE CREATION OF ELIZA AS ARTIFACT AND NARRATIVE

Although the implications and actual meaning of the Turing test are still at the center of a lively debate, it is beyond doubt that the test contributed to setting practical goals for the community of AI researchers that developed in the following decades. In what is probably the most influential AI handbook in computer science, Stuart Russell and Peter Norvig recognize the key role of the test in the development of a particular understanding of research in this field based on the so-called behavioral approach, which pursues the goal of creating computers that *act* like humans. Programs designed in behavioral AI, they explain, are designed to exhibit rather than actually replicate intelligence, thereby putting aside the problem of what happens inside the machine's "brain."[3]

As Weizenbaum took up an academic position at MIT in 1964 to work on AI research, his work was informed by a similar approach.[4] In his writings, he professed that AI was and should be distinguished from human intelligence. Yet he made efforts to design machines that could lead people into believing they were interacting with intelligent agents, since he was confident that users' realization that they had been deceived would help them understand the difference between human intelligence and AI.

Between 1964 and 1966, Weizenbaum created what is considered the first functional chatbot, that is, a computer program able to interact with users via a natural language interface.[5] The functioning of ELIZA was rather simple. As Weizenbaum explained in the article describing his invention, ELIZA searched the text submitted by its conversation partner for relevant keywords. When a keyword or pattern was found, the program produced an appropriate response according to specific transformation rules. These rules were based on a two-stage process by which the input was first decomposed, breaking down the sentence into small segments. The segments were then reassembled, readapted according to appropriate rules—for instance by substituting the pronoun "I" for "you" —and programmed words were added to produce a response. In cases when it was impossible to recognize a keyword, the chatbot would employ preconfigured formulas, such as "I see" or "Please go on" or alternatively would create a response through a "memory" structure that drew from previously inserted inputs.[6]

Made available to users of the Project MAC time-sharing system at MIT, the program was designed to engage in conversations with human users who responded by writing on a keyboard, a situation similar to a contemporary messaging service or online chatroom. In this sense, the very possibility of ELIZA resided on the availability of a community of users who could engage in conversations with the chatbot thanks to the development of new human-computer interaction systems.[7]

Weizenbaum was adamant in his contention that ELIZA exhibited not intelligence but the illusion of it. The program would demonstrate that humans in interactions with computers were vulnerable to deception.[8] As Weizenbaum conceded in an interview with Daniel Crevier, a historian of AI, ELIZA was the immediate successor of a program that played a game called Five-in-a-row or Go-MOKU, which was described in his first published article, aptly titled "How to Make a Computer *Appear* Intelligent."[9] The program used a simple strategy with no lookahead, yet it could beat anyone who played at the same naive level and aimed at creating "a powerful illusion that the computer was intelligent."[10] As noted in the article where it was first described, it was able to "fool some observers for some time." Indeed, deception was, to Weizenbaum, the measure of success for the author of an AI program: "his success can be measured by the percentage of the exposed observers who have been fooled multiplied by the length of time they have failed to catch on." On the basis of this criterion, Weizenbaum considered his program to be quite successful, as it made many observers believe that the computer behaved intelligently, providing a "wonderful illusion of spontaneity."[11]

Considering the limited computer power and resources available, ELIZA was also quite successful in deceiving users, at least when the interaction was limited to a relatively brief conversation. Its efficacy was due to some intuitions that did not strictly pertain to the domain of programming but derived from insights from psychology and from Weizenbaum's understanding of human behavior in conversations. He had realized that our perception of the identity of a conversation partner is crucial to the credibility of any human interaction. Thus, in order to pass convincingly for a human, a chatbot should not only respond correctly to a given input but also play a coherent role throughout the conversation.[12] Consequently, he conceived of ELIZA as a program that could be adapted to different roles, which he called, using one of his characteristic theatrical metaphors, *scripts*. In ELIZA's software architecture, scripts were treated as data, which implied that they were "not part of the program itself." In terms of conversation patterns, a script corresponded to a specific part that the bot would play throughout a conversation.[13]

In the initial version of ELIZA, called DOCTOR, the program's script simulated a psychotherapist employing the Rogerian method—a type of nondirective therapy in which the therapist reacts to the patient's talk mainly by redirecting it back to the patient, often in the form of further questions.[14] The choice of this role was crucial to ELIZA's success: in fact, the dynamics of the therapy allowed the program to sustain a conversation while adding little, if anything, to it. The results are evident in some excerpts of a conversation with ELIZA that Weizenbaum published in his first article on the subject. (ELIZA's contributions are in capital letters.)

> Men are all alike.
> IN WHAT WAY
> They're always bugging us about something or other.
> CAN YOU THINK OF A SPECIFIC EXAMPLE
> Well, my boyfriend made me come here.
> YOUR BOYFRIEND MADE YOU COME HERE
> He says I'm depressed much of the time.
> I AM SORRY TO HEAR YOU ARE DEPRESSED.[15]

The name ELIZA was based, Weizenbaum explained, on the character Eliza Doolittle in George Bernard Shaw's play *Pygmalion*. In the play, phonetics specialist Professor Higgins makes a bet with a friend that with his knowledge of phonetics he can train Eliza, a Covent Garden florist with a cockney accent, into a woman as poised and well-spoken as a member of the aristocracy.[16] This corresponded to Weizenbaum's hope that his program would become more refined and varied, just like the salesgirl under the tutelage of Professor Higgins.[17] The choice of a literary work such as *Pygmalion*, which interrogates the issue of authenticity and performance, was not accidental. Faithful to Weizenbaum's behavioral approach, which, as explained earlier, aimed at simulating rather than replicating human intelligence, ELIZA resembled the Eliza of *Pygmalion* fame because it created an appearance of reality, remaining, however, "in the domain of the playwright."[18] In fact, Weizenbaum often compared his creation to an actor or actress who "improvises according to certain guidelines, that is within a certain system or—let's keep the image of the theatre—performs a certain role."[19] In order to underline that ELIZA was only able to produce the impression of reality, he went so far as to point out that "in a sense ELIZA was an actress who commanded a set of techniques but who had nothing of her own to say" and to describe it as a "parody" of a nondirective psychotherapist.[20]

To Weizenbaum, then, ELIZA's operations could be equated to acting, and more broadly, conversation—and consequently, verbal human-machine interactions—was a matter of role playing. As he made clear in an article dedicated to the problem of, as the title put it, "Contextual Understanding by Computers," he realized that the contributions of human subjects were central to such interactions. Humans make assumption about their conversation partners, assigning them specific roles; thus, a conversation program is credible as long as it successfully plays its part. The illusion produced by the program will break down in the moment when the contextual assumptions made by the human partner cease to be valid—a phenomenon that, Weizenbaum notes, is at the basis of the comical effects created in comedies of errors.[21]

In the decades since the creation of ELIZA, the theatrical metaphor has been used by scholars and commentators to discuss Weizenbaum's work and, more broadly, has become a common way commentators and developers alike describe the functioning of chatbots.[22] This is significant because, as George Lakoff and Mark Johnson famously demonstrated, the emergence of new metaphors to describe things and events may result in orienting attitudes toward them and in guiding future actions.[23] While the metaphor of the theatre was used by Weizenbaum to demonstrate that AI was the fruit of an illusory effect, another comparison he employed pointed even more explicitly to the issue of deception: he noted, in fact, that users' belief that ELIZA was actually understanding what they were saying "is comparable to the conviction many people have that fortune-tellers really do have some deep insight."[24] Like a fortune teller, ELIZA's messages left enough space of interpretation for users to fill the gaps, so that "the 'sense' and the continuity the person conversing with ELIZA perceives is supplied largely by the person himself."[25]

As historians of technology have shown, technologies function not only at a material and technical level but also through the narratives they generate or into which they are forced.[26] David Edgerton, for instance, points to the V2 rockets the German army developed and employed at the end of World War II. Although the rockets were ineffective if one considers the amount of resources used and their practical effects, their development was useful at a symbolic level, since it kept the hopes of victory alive. The rockets functioned, in other words, first and foremost, at a narrative rather than a material level.[27] A similar discourse could apply to software. Think, for instance, of the Deep Blue chess program, which famously beat chess master Garry Kasparov in 1997. IBM had made considerable investments to develop Deep Blue, yet once the challenge was won and the attention of journalists evaporated, IBM dismantled the project and reassigned the engineers working on it to different tasks. The company, in fact, was interested in the program's capacity to create a narrative—one that would place IBM at the forefront of progress in computation—more than in its potential application beyond the Kasparov challenge.[28]

In a similar way, ELIZA was especially effective at a discursive rather than a practical level. Weizenbaum was always transparent about the fact that ELIZA had limited practical application and would be influential in shedding light on a potential path rather than in the context of its immediate use.[29] As Margaret Boden rightly points out, in terms of programming work ELIZA was simple to the point of being obsolete even at the moment of its creation, and proved largely irrelevant in technical terms for the development of the field, essentially because Weizenbaum "wasn't aiming to make a computer 'understand' language."[30] His deep interest, on the contrary, in how the program would be interpreted and "read" by users suggests that he aimed at the creation of an artifact that could produce a specific narrative about computers, AI, and the interaction between humans and machines. ELIZA, in this regard, was an artifact created to prove his point that AI should be understood as an effect of users' tendency to project identity. In other words, ELIZA was a narrative about conversational programs as much as it was a conversational program itself.

Weizenbaum, however, was not interested in deceiving computer users. On the contrary, he expected that a particular interpretation of ELIZA would emerge from users' interactions with it, as they realized that the apparent intelligence of the machine was just the result of deception.[31] This narrative would present AI not as the result of humanlike intelligence programmed into the machine but as an illusory effect. This narrative would therefore replace the myth of the "thinking machine," which suggested that computers could equal human intelligence, with a narrative more consistent with the behavioral approach in AI.[32] Looking back at ELIZA's creation, he explained that it was conceived as a "programming trick"[33]—even a "joke."[34] By showing that a computer program of such limited complexity could trick humans into believing it was real, ELIZA would work as a demonstration of the fact that humans, facing "AI" technologies, are vulnerable to deception.

Weizenbaum's approach recalls in this sense the use in the Victorian age of philosophical toys such as the kaleidoscope or the phenakistoscope, which illustrated an idea about optics through a device that manipulated the viewers' perceptions and at the same time entertained them.[35] In fact, Weizenbaum noted that one of the reasons for ELIZA's success was that users could interact playfully with it.[36] It was targeted to users of the new time-sharing system at MIT, who would be stimulated by it to reflect not only on AI's potentials but also, and crucially, on its limitations, since the "programming trick" created the illusion of intelligence rather than intelligence itself.

Weizenbaum's success in turning ELIZA into the source of a specific narrative about AI as deception is evident in the stories that circulated about ELIZA's reception. One famous anecdote concerned Weizenbaum's secretary, who once asked him to leave the room, needing some privacy to chat with ELIZA. He was particularly startled by this request because the secretary was well aware of how the program functioned and could hardly consider it a good listener.[37] Another anecdote about ELIZA concerns a computer salesman who had a teletype interchange with ELIZA without being aware that it was a computer program; the interaction resulted in him losing his temper and reacting with fury.[38] Both anecdotes have been recalled extremely often to stress humans' tendency to be deceived by AI's appearance of intelligence—although some, like Sherry Turkle, point out that the story of the secretary might reveal instead users' tendency to maintain the illusion that ELIZA is intelligent "because of their own desires to breathe life into a machine."[39]

In such anecdotes, which played a key role in informing the program's reception, the idea that AI is the result of deception becomes substantiated through a simple, effective narrative. Indeed, one of the characteristics of anecdotes is their capacity to be remembered, retold, and disseminated, conveying meanings or claims about the person or thing they refer to. In biographies and autobiographies, for instance, anecdotes add to the narrative character of the genre, which despite being nonfictional is based on storytelling, and at the same time contribute to the enforcing of claims about the person who is the subject of the biographical sketch, for example, her temperament, personality, and skills.[40] In the reception of ELIZA, the anecdote about the secretary played a similar role, imposing a recurring pattern in which the functioning of the program was presented to the public in terms of deception. Julia Sonnevend has convincingly demonstrated that one of the characteristics of the most influential narratives about media events is their "condensation" into a single phrase and a short narrative.[41] Others have referred to a similar process in Latourian terms as a "pasteurization" of the narrative, in which "germs are eliminated in the name of a simple, rational and powerful explanation."[42] Through the pasteurization of the narrative, elements that do not fit with the dominant narrative about a given event are disregarded, privileging a more coherent and stable narrative. In this sense, the anecdotes about deception and particularly the story of the secretary "condensed" or "pasteurized" the story of ELIZA, presenting it to the public as a powerful narrative about computers' capacity to deceive users.

THE COMPUTER METAPHOR AND THE NARRATIVE OF THINKING MACHINES

As I have shown, Weizenbaum's writings demonstrate that he also programmed ELIZA specifically with the intention to present a particular vision of AI based on what can be described as the narrative of deception. This was substantiated in anecdotes about the reception of ELIZA, which circulated widely and shaped discussions about the program's meanings and implications. Weizenbaum did not consider, however, that while he was able to control the program's behavior, he would not be able to control—or to use computer science language, to "program"—all the narratives and interpretations that would emerge from it. Yet, unexpectedly for its creator, the reception of ELIZA also entailed the emergence of a second narrative, which presented the program's proficiency not as a successful illusion but instead as evidence that computers can equate or surpass human intelligence.

As discussed in chapter 2, computers from their early history have been presented as mechanical or electronic brains whose operations might be able to replicate and surpass human reason.[43] In narrative form, this vision corresponds to stories in a wide range of science fiction scenarios in which robots and computers display human characteristics, as well as to journalistic reports that exaggerate computers' achievements.[44] Although results in AI, especially in the early history of the field, were often far from being comparable to human intelligence, even researchers who were aware of the limits of the field tended to overstate AI's achievements, nurturing the narrative of computers as "thinking machines."[45]

This narrative contrasts sharply with Weizenbaum's tenet that AI should be understood in terms of an illusory effect rather than as evidence that the machine understands and reasons like humans. He contended that believing that computers were thinking machines was similar to entertaining superstitious beliefs. In his first article describing ELIZA, he pointed out that computers might be regarded by laypersons as though they are performing magic. However, "once a particular program is unmasked, once its inner workings are explained in language sufficiently plain to induce understanding, its magic crumbles away; it stands revealed as a mere collection of producers, each quite comprehensible."[46] By making this point, Weizenbaum seemed unaware of a circumstance that concerns any author—from the writer of a novel to an engineer with her latest project: once one's creation reaches public view, no matter how carefully one has reflected on the meanings of its creation, these meanings can be overturned by the readings and interpretations of other writers, scientists, journalists, and laypersons. A computer program can look, despite the programmer's intention, as if it is performing magic; new narratives may emerge, overwriting the narrative the programmer meant to embed in the machine.

This was something Weizenbaum would learn from experience. The public reception of ELIZA, in fact, involved the emergence a very different narrative from the one he had intended to "program" into the machine. With the appearance in 1968 of Stanley Kubrick's now classic science fiction film *2001: A Space Odyssey*, many thought that ELIZA was "something close to the fictional HAL: a computer program intelligent enough to understand and produce arbitrary human language."[47] Moreover, Weizenbaum realized that research drawing or following from his work was led by very different understandings about the scope and goals of AI. A psychologist from Stanford University, Kenneth Mark Colby, developed PARRY, a conversational bot whose design was loosely based on ELIZA but represented a very different interpretation of the technology. Colby hoped that chatbots would provide a practical therapeutic tool by which "several hundred patients an hour could be handled by a computer system designed for this purpose."[48] In previous years, Weizenbaum and Colby had collaborated and engaged in discussions, and Weizenbaum expressed later some concerns that his former collaborator did not give appropriate credit to his work on ELIZA; but the main issue in the controversy that ensued between the two scientists was on moral grounds.[49] A chatbot providing therapy to real patients was in fact a prospect Weizenbaum found dehumanizing and disrespectful of patients' emotional and intellectual involvement, as he later made abundantly clear.[50] The question arises, he contended, "do we wish to encourage people to lead their lives on the basis of patent fraud, charlatanism, and unreality? And, more importantly, do we really believe that it helps people living in our already overly machine-like world to prefer the therapy administered by machines to that given by other people?"[51] This reflected his firm belief that there were tasks that, even if theoretically or practically possible, a computer should not be programmed to do.[52]

ELIZA attracted considerable attention in the fields of computer science and AI, as well as in popular newspapers and magazines.[53] These accounts, however, often disregarded the fact that its effectiveness was due to a deceptive effect. On the contrary, ELIZA's success in looking like a sentient agent was presented in a way that supported widespread narratives about computers as "thinking machines," thereby exaggerating the capabilities of AI. This presentation was precisely what Weizenbaum had hoped to avoid since his earliest contributions to the field, in which he had intended to unveil what makes a computer *appear* intelligent, debunking an impression of intelligence and understanding that he considered misleading. This prompted him to redesign ELIZA so that the program revealed its misunderstandings, provided explanations to users, and dispelled their confusion.[54]

Weizenbaum worried that the ways the functioning of ELIZA was narrated by other scientists and in the press contributed to strengthening what he called the "computer metaphor," by which machines were compared to humans, and software's capacity to create the appearance of intelligence was exchanged for intelligence itself.[55] In numerous writings over the years, Weizenbaum would lament the tension between reality and the public perception of the actual functioning of computers. The problem, he pointed out, was not only the gullibility of the public; it was also the tendency of scientists to describe their inventions in exaggerated or inaccurate ways, profiting from the fact that nonexperts are unable to distinguish between what might and might not be true. In a letter to the *New York Times* in which he criticized the overoptimistic claims made by Jeremy Bernstein in an article on self-reproducing machines, Weizenbaum pointed to the "very special responsibility" of the science writer to the lay reader who "has no recourse but to interpret science writing very literally."[56] As Weizenbaum noted in a conversation with a journalist at the *Observer*, experts' narratives could have practical effects, since "language itself becomes merely a tool for manipulating the world, no different in principle from the languages used for manipulating computers."[57] In other words, to Weizenbaum, the use of particular discourses and narratives could have effects on reality much in the way that the instructions coded by a programmer might result in operations that trigger changes in the real world.[58] For this reason Weizenbaum regarded it as an important duty for scientists to be accurate and avoid sensationalism when communicating their research.

In *Computer Power and Human Reason* (1976), Weizenbaum points out that laypeople are especially vulnerable to distorting representations in this context, because computers are technologies whose actual functioning is little known but that exercise a strong power on the imagination of the public, bringing about "a certain aura—derived, of course, from science."[59] He worried that if public conceptions of computing technologies were misguided, then public decisions about the governance of these technologies were likely to be misguided as well.[60] The fact that the computer had become a powerful metaphor, in his view, did nothing to improve understandings of these technologies among the public. On the contrary, he reasons, a metaphor "suggests the belief that everything that needs to be known is known," thereby resulting in a "premature closure of ideas" that sustains rather than mitigates the lack of understanding of the complex scientific concepts related to AI and computation.[61]

Scholars of digital culture have shown how metaphors such as the cloud provide narratives underpinning powerful, and sometimes inaccurate, representations of new technologies.[62] Less attention has been given, however, to the role of particular artifacts, such as software, in stimulating the emergence of such metaphors. Yet the introduction of software often stimulates powerful reactions in the public sphere. Digital games such as *Death Race* or *Grand Theft Auto*, for instance, played a key role in directing media narratives about the effects of gaming.[63] As Weizenbaum perceptively recognized, the ways ELIZA's functioning was narrated by other scientists and in the press contributed to strengthening the "computer metaphor," by which software's capacity to create the appearance of intelligence was exchanged for intelligence itself. Although he had conceived and programmed ELIZA with the expectation that it would help people dismiss the magical aura of computers, the narratives that emerged from public scrutiny of his invention were actually reinforcing the very same metaphors he intended to dispel.

REINCARNATING ELIZA

Apple's developers programmed Siri to respond to queries about ELIZA with the sentence "ELIZA is my good friend. She was a brilliant psychologist, but she's retired now" (fig. 3.1). In doing so, they aimed not only to give homage to the first exemplar of a kind but also to make their own creation, Siri, appear smart. One of the ways Siri as a character differs from other voice assistants, such as Alexa, is in fact Siri's use of irony, through which it is given a distinctive personality. On the Internet, one can find several web pages reporting some of the funniest answers from Siri. If asked to "talk dirty," for instance, Siri (or better said, some versions of Siri) responds: "The carpet needs vacuuming."[64] Similarly, the particular wording of Siri's sentence about ELIZA also displays a degree of irony. The link between the two systems is presented as a friendship, a kind of affective relationship that contrasts with their nature as artifacts yet reveals the developers' ambition to create the illusion of humanity. ELIZA is described as a *brilliant* psychiatrist, even though Weizenbaum incessantly emphasized that it provided only the illusion of intelligence and was, in fact, the opposite of clever. In referring to ELIZA's retirement, moreover, Siri's developers jokingly play with the fact that it was part of another generation, whose legacy has been taken up by virtual assistants such as Siri and the like.

Indeed, irony might be the best way to pay homage to the first chatbot in history. As discussed earlier, Weizenbaum stressed numerous times that ELIZA was meant to be a parody of a psychotherapist, underlining the ironic intention of his creation. It was a joke about psychotherapy, the caricature of a psychiatrist. As Janet H. Murray points out in her study on how digital media are reshaping narrative, ELIZA was also "the first completely computer-based character" that gave "parodic emphasis to some familiar absurdities of therapy itself."[65] If fictional characters from literature, cinema, or television are sometimes so familiar to the public that people treat them as if they exist for real, something similar has happened with the memory of ELIZA, which became an electronic specter haunting contemporary AI systems and, more broadly, computer culture on the whole.

The legacy of Weizenbaum's creation is still present today not in one way but in multiple ways. As Noah Wardrip-Fruin notes, ELIZA became in the decades after its creation "one of the world's most famous demonstrations of the potential of computing."[66] ELIZA is discussed as an experiment posing key questions about how users project sense and meaning onto machines. And ELIZA also continues to interact with contemporary computer users, as a chatbot that anyone equipped with a device and an Internet connection can start a dialogue with. In fact, several reincarnations of ELIZA are available for users to chat with at different websites on the Internet (fig. 3.2). They are avatars of Weizenbaum's creation, though not exact copies, as they are rewritten in a different programming language.[67] The way contemporary users are invited to engage playfully with them reinforces the comparison with philosophical toys of the Victorian age, which served to prove a point about illusion and perception as well as to entertain. In a similar way, ELIZA's avatars encourage playful interactions online, but also stimulate reflections on the nature of AI systems and on people's inner drive to enter into relationship with them.

The high media profile of ELIZA was instrumental, as Boden points out, to bring the Turing test to the vocabulary of laypersons as well as to reinforce its presence in the vocabulary of AI.[68] Especially at a discursive level, the legacy of ELIZA has played a crucial role in shaping reflections

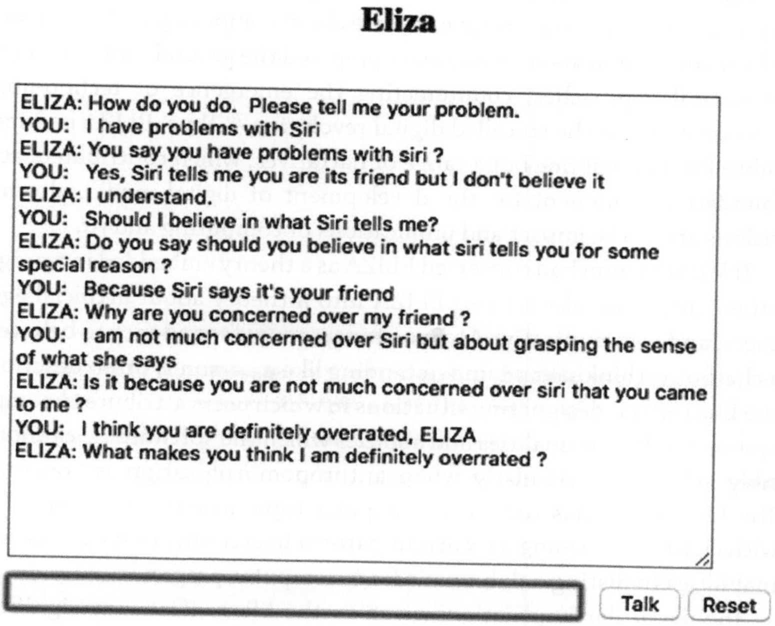

Eliza

ELIZA: How do you do. Please tell me your problem.
YOU: I have problems with Siri
ELIZA: You say you have problems with siri ?
YOU: Yes, Siri tells me you are its friend but I don't believe it
ELIZA: I understand.
YOU: Should I believe in what Siri tells me?
ELIZA: Do you say should you believe in what siri tells you for some special reason ?
YOU: Because Siri says it's your friend
ELIZA: Why are you concerned over my friend ?
YOU: I am not much concerned over Siri but about grasping the sense of what she says
ELIZA: Is it because you are not much concerned over siri that you came to me ?
YOU: I think you are definitely overrated, ELIZA
ELIZA: What makes you think I am definitely overrated ?

Talk Reset

Figure 3.2 Author's conversation with ELIZA's avatar, at http://www.masswerk.at/elizabot/, 18 May 2018. The author acknowledges to have himself fallen into the (banal?) deception, by referring to a piece of software through a feminine instead of a neutral pronoun.

and practical experiments in making humans and machines interact with each other through language up to the present day. Mark Marino's doctoral dissertation presents a survey provides evidence that communities of chatbot users and makers continue to consider ELIZA "the progenitor, a gold standard, and the relic, a significant touchstone in current evaluations of chatbots."[69]

One of ELIZA's peculiarities is that it was conceived from the start as embodying a specific narrative about computing and its relationship with humans. As I have shown, Weizenbaum regarded the program as an artifact illustrating a theory about AI and human-computer interactions. This, however, proved to be only one of the interpretations associated with ELIZA. In fact, narratives that presented the program as evidence of computers' intelligence—the computer metaphor—shaped the public presence of the software. This in turn stimulated Weizenbaum to propose a different interpretation of the event, which supported his call to reflect critically on the dangers and potential problems related to the use of computing technologies. His widely read book *Computer Power and Human Reason* turned ELIZA into a powerful narrative about computing and digital media, which anticipated and in many ways prepared the ground for the emergence of critical approaches, counteracting the emergence of techno-utopian discourses about the so-called digital revolution.[70] Thus, ELIZA became the subject of not just one but a range of narratives, which contributed, at that foundational moment for the development of digital media, alternative visions about the impact and implications of computing and AI.

If Weizenbaum had conceived ELIZA as a theory embedded in a program, others after him also turned ELIZA into a theory about artificial agents, users, and communicative AI. Even today, people's tendency to believe that a chatbot is thinking and understanding like a person is often described as the Eliza effect, designating situations in which users attribute to computer systems "intrinsic qualities and abilities which the software . . . cannot possibly achieve," particularly when anthropomorphization is concerned.[71] The Eliza effect has remained a popular topic in reports of interactions with chatbots, forming a recurrent pattern in accounts of users' occasional inabilities to distinguish humans from computer programs.

The particular implications given to the Eliza effect vary significantly from one author to the other. Noah Wardrip-Fruin, for instance, suggests that the deception was the result of audiences developing a mistaken idea about the internal functioning of the chatbot: "they assumed that since the surface appearance of an interaction with the program could resemble something like a coherent dialogue, internally the software must be complex."[72] The Eliza effect, in this sense, designates the gap between the black-boxed mechanisms of software and the impression of intelligence that users experience. This impression is informed by their limited access to the machine but also by their preconceptions about AI and computers.[73]

Turkle gives a different and in many respects more nuanced interpretation of the Eliza effect. She uses this interpretation to describe the ways users, once they knew that ELIZA or another chatbot is a computer program, change their modes of interaction so as to "help" the bot produce comprehensible responses. She notes, in fact, that ELIZA users avoid saying things they think will confuse it or result in too predictable responses. In this sense, the Eliza effect demonstrates not so much AI's capacity to deceive as the tendency of users' to fall willingly—or perhaps most aptly, complacently—into the illusion.[74]

This interpretation of the Eliza effect recalls wider responses that people have to digital technologies, if not toward technology on the whole. As Jaron Lanier observes, "we have repeatedly demonstrated our species' bottomless ability to lower out our standards to make information technology look good": think, for instance, of teachers giving students standardized tests that an algorithm could easily solve, or the bankers before the financial crash who continued to trust machines in directing the financial system.[75] The Eliza effect, therefore, reveals a deep willingness to see machines as intelligent, which informs narratives and discourses about AI and computing as well as everyday interactions with algorithms and information technologies.

In this sense ELIZA, as Turkle put it, "was fascinating not only because it was lifelike but because it made people aware of their own desires to breathe life into a machine."[76] Although this desire usually remains tacit in actual interactions with AI, it may reveal also a form of pragmatism on the part of the user. For example, if a chatbot is used to do an automated psychotherapy session, falling under the spell of the Eliza effect and thus "helping" the chatbot do its job could potentially improve the experience of the user. Similarly, AI companions—such as social robots or chatbots—that are designed to keep users company will be more consoling if users prevent them from revealing their shortcomings by choosing inputs they can reply to. In addition, users of AI voice assistants such as Siri and Alexa learn the most appropriate wordings for their queries, to ensure that these assistants will perform their daily tasks and "understand" them. As a manifestation of banal deception, in this sense, the Eliza effect contributes to explaining why users are complicit with the deceptive mechanisms embedded in AI: not as a consequence of their naivete but rather because the deception has some practical value to them.

Still, such apparent benevolence should not make us forget Weizenbaum's passionate call to consider the benefits as much as the risks of AI technologies. The Eliza effect, to Weizenbaum himself, epitomized a dynamics of reception that was "symptomatic of deeper problems": people were eager to ascribe intelligence even if there was little to warrant such a view.[77] This tendency could have lasting effects, he reasoned, because humans, having initially built the machines, might come to embrace the very models of reality they programmed into them, thereby eroding what we understand as human. As such, for Weizenbaum ELIZA demonstrated that AI was not or at least not only the technological panacea that had been enthusiastically heralded by leading researchers in the field.

OF CHATBOTS AND HUMANS

Taina Bucher recently proposed the notion of the "algorithmic imaginary" to describe the ways people experience and make sense of their interactions with algorithms in everyday life.[78] She notes that everyday encounters with computing technologies are interpreted in personal ways by users, and that these interpretations inform how they imagine, perceive, and experience digital media. While Bucher focuses mainly on the context of interaction with modern computing technologies, the case of ELIZA shows that these interpretations are also informed by the narratives and visions that circulate around them. In fact, as argued by Zdenek, AI depends on the production of physical as well as discursive artifacts: AI systems are "mediated by rhetoric" because "language gives meaning, value and form" to them.[79] This dependence is also due to the characteristic opacity of software, whose functioning is often obscure to users and, to a certain extent, even to computer scientists, who often work with only a partial understanding and knowledge of complex systems. If narratives in this sense may illuminate software's technical nature for the public, they might at the same time turn software into contested objects whose meanings and interpretations are the subject of complex negotiations in the public sphere.

ELIZA was one such contested object and, as such, prompted questions about the relationship between AI and deception that resonate to the present day. In the context of an emerging AI field, ELIZA provided an insight into patterns of interaction between humans and computers that only apparently contrasts with the vision of human-computer symbiosis that was embraced at the same time at MIT and in other research centers across the world. In fact, Weizenbaum's approach to AI in terms of illusion was not by any means irrelevant to the broader discussions about human-machine interactions that were shaping this young discipline.

In histories of computing and AI, Weizenbaum is usually presented as an outcast and a relatively isolated figure within the AI field. This was certainly true during the last part of his career, when his critical stance gained him the reputation of a "heretic" in the discipline. This was not the case, however, during his early work on ELIZA and other AI systems, which were extremely influential in the field and secured him a tenured position at MIT.[80] The questions Weizenbaum was posing at the time of his work on ELIZA were not the solitary explorations of an outcast; several of his peers, including some of the leading researchers of the time (as chapter 2 has shown), were asking them too. While AI developed into a heterogeneous milieu bringing together multiple disciplinary perspectives and approaches, many acknowledged that users could be deceived in interactions with "intelligent" machines. Like Weizenbaum, other researchers in the burgeoning AI field realized that humans were not an irrelevant variable in the AI equation: they actively contributed to the emergence of intelligence, or better said, the appearance of it.

The questions that animated Weizenbaum's explorations and the controversy about ELIZA, in this sense, informed not only the work of chatbot developers and AI scientists. In different areas of applications, software developers discovered that users' propensity to project intelligence or agency onto machines was kindled by specific elements of software's design, and that mobilizing such elements could help achieve more functional programs and systems. In the next chapter, I explore some examples of ways the creation of software turned into a new occasion to interrogate how deception informed the interaction between computer and users. For all their specificity, these examples will show that the Eliza effect concerns much more than an old chatbot that might have deceived some of its users: it is, on the contrary, a structural dimension of the relationship between humans and computing systems.

CHAPTER 4

Of Daemons, Dogs, and Trees

Situating Artificial Intelligence in Software

Between the 1970s and the early 1990s, in coincidence with the emergence of personal computing and the shift to network systems and distributed computing, AI took further steps toward being embedded in increasingly complex social environments.[1] AI technologies were now more than isolated experiments testing the limits and potentials of computing, as had been largely the case until then. More and more often these technologies were made available to communities of users and aimed at practical applications. Pioneering projects like ELIZA had mostly concerned communities of practitioners such as the one surrounding the MAC time-sharing project at MIT. But when computing devices moved to the center of everyday life for masses of people, the significance of AI changed irremediably. So too did its meaning: AI took up a social life of its own, as new opportunities emerged for interactions between artificial agents and human users.

Somewhat paradoxically, these developments coincided with a lack of fortune for the AI enterprise. After the enthusiasm of the 1950s and 1960s, a period of disillusion about the prospects of AI characterized the following two decades. The gap between the visionary myths of thinking machines and the actual outcomes of AI research became manifest, as a series of critical works underlined the practical limitations of existing systems. The Lighthill report, commissioned by the British Science Research Council, for instance, gave a pessimistic evaluation of the drawbacks of AI, identifying the "combinatorial explosion": in increasing the size of the data, some

Deceitful Media. Simone Natale, Oxford University Press (2021). © Oxford University Press.
DOI: 10.1093/oso/9780190080365.003.0005

problems become rapidly intractable.[2] Other critical works, such as Herbert Dreyfus's *Alchemy and Artificial Intelligence*, also highlighted that the much-heralded achievements of AI research were far from bringing significant practical results.[3] The criticism seriously undermined the AI community as a whole, resulting in a climate of public distrust regarding their research and in a loss of credibility and funding: the so-called AI winter.

As a response to this, many researchers in the late 1970s and in the early 1980s started to avoid using the term *AI* to label their research even when working on projects that would have been called AI just a few years earlier. They feared that if they adopted such a term to describe their work, the general lack of credibility of AI research would penalize them.[4] Even if labeled otherwise, however, research into areas such as natural language processing and communicative AI continued to be done in laboratories across the United States, Europe, Asia, and the world, with significant achievements.[5] In addition, the rising computer software industry supported experimentation with new ways to apply AI technologies in areas like home computing and gaming. During this time, crucial steps were taken that implemented AI within new interactive systems and prepared for today's communicative AI—the likes of Alexa, Siri, and Google Assistant.

This chapter examines the ways AI was incorporated in software systems by focusing on three particular cases: routine programs called daemons, digital games, and social interfaces. The trajectory of these different software artifacts provides an opportunity to reflect on the way users' understandings of computers and AI have changed over time, and how such changes are connected with evolutions in interactions between humans and computing machines. Observing a significant change since the late 1970s to the 1990s in how participants in her studies perceived AI, Turkle has suggested that the emergence of personal computing and the increasing familiarity with computing technologies made AI appear less threatening to users. As computers became ubiquitous and took up an increasing number of tasks, their inclusion in domestic environments and everyday lives made them less frightening and more familiar to people.[6] The integration of AI-based technologies in software that came to be in everyday use for millions of people around the world helped users and developers to explore the potentials and ambiguities of AI, preparing for the development of contemporary AI systems, not only at a technical but also at a cultural and social level.

Reconstructing histories of software, as of all technical artifacts, requires a multilayered approach that embraces the material, social, and cultural dimensions of technology.[7] Yet the peculiar character of computer software adds extra levels of complexity to this enterprise. What makes the history of software hard, as historian of computing Michael S. Mahoney points out, is that it is not primarily about computers. Software reflects on one side the histories of the communities that created them and on the other the cultural, social, and practical circumstances underpinning its adoptions and uses. It is in the specific situations and contexts in which software is used and circulated that one can trace its actual material outcomes and functioning.[8] Tracing histories of software, therefore, requires a perspective that is sensitive to its manifold dimensions.[9] It is in this spirit that I turn to the histories of software artifacts as diverse as computer daemons, digital games, and social interfaces. Each of the objects under examination tells us something about what AI means today, as opportunities for integrating AI into everyday lives and environments continue to expand.

TALKING TO STONES: DAEMONS, SOFTWARE AGENCY, AND THE IMAGINATION

A long-standing debate in computer science revolves around the question of computers "creating" beyond what their programmers intended or expected. The discussion goes back to the mathematician who is credited with being the first computer programmer in history, Ada Lovelace, who understood as early as the mid-nineteenth century that calculators could be used not only to compute numbers but also to compose music, produce graphics, and advance science.[10] While the issue has been debated passionately, today few deny software's capacity to transcend programmer agency. In fact, due to the complexity of contemporary software systems and the opacity of the operations performed by neural networks, programmers often cannot anticipate or even understand software output. But there is an additional reason why software cannot be reduced to the agency of programmers. Once "out there," software is adapted to different platforms, situated in different contexts, embraced by different users and communities, and "performed" in concrete, temporal interactions.[11] Software is applied or recycled to fulfill very different goals from those originally intended. It is integrated into social and cultural environments different from those of its origin. It is described, perceived, and represented through cultural constructions that also change throughout time.

That's why apparently obscure and ordinary lines of software often generate complex dynamics of interpretation and meaning. One of the best examples of this are the so-called daemons, that is, computer programs that run in the background and perform tasks without the direct intervention of the user. Daemons might be embedded in a system or an environment but run independently from other processes. Once active, they follow the coded instruction to complete routine tasks or react to particular situations or events. A crucial distinction of daemons from other software utilities is that they do not involve interaction with users: "a true daemon," as Andrew Leonard put it, "can act on its own accord."[12] Daemons deliver vital services enabling the smooth functioning of operating systems, computer networks, and other software infrastructures. Email users may have encountered the pervasive mailer-daemon that responds to various mailing problems, such as sending failures, forwarding needs, and lack of mailbox space. Although unbeknownst to them, navigators visiting any web page also benefit from the daemons that respond to requests for access to web documents.[13]

The origins of the word *daemon* are noteworthy. In Greek mythology, daemons were both benevolent and evil divinities. A daemon was often presented as an intermediary between gods and mortals—a *medium* in the literal sense, that is, "what is in between." In nineteenth-century science, the word *daemon* came to be employed for describing fictional beings that were included in thought experiments in domains like physics. These daemons helped scientists picture situations that could not be reproduced in practical experiments but were relevant to theory.[14] The most famous of these daemons was the protagonist of a thought experiment by British scientist James Clerk Maxwell in 1867, wherein he suggested the possibility that the second law of thermodynamics might be violated. According to Fernando Corbato, who is usually credited with inventing computer daemons, "our use of the word daemon was inspired by the Maxwell's daemon of physics and thermodynamics. . . . Maxwell's daemon was an imaginary agent which helped sort molecules to different speeds and worked tirelessly in the background. We fancifully began to use the word daemon to describe background processes which worked tirelessly to perform system chores."[15]

In human-computer interaction, daemons contribute to making computers "transparent" by staying in the background and performing tasks that help the interface and the underlying system work efficiently and seamlessly. Thus, daemons are often characterized as servants who anticipate the master's wishes and use their own initiative to meet such needs, without being noticed by the master-user.[16] As Fenwick Mckelvey recently noted, however, daemons are never mere servant mechanisms. Given their autonomy and the role they play in large digital infrastructures such as the Internet, they should also be seen as mechanisms controlling these infrastructures: "since daemons decide how to assign and use finite network resources, their choice influence which networks succeed and which fail, subtly allocating resources while appearing to maintain the internet's openness and diversity."[17]

Despite their crucial role in deciding the shape of the Internet and other systems, one should resist attributing intelligence or personalities to daemons. Not differently from other software programs, daemons are made up of lines of code. In our eyes, it is their capacity to act autonomously that distinguishes them from software such as word processors. This can be seen as agency similarly to the way things—as social anthropologists Arjun Appadurai and Alfred Gell have taught us—can be regarded as social agents.[18] One can easily observe, in fact, that people attribute intentions to objects and machines—think, for instance, at one's anger toward a malfunctioning television or PC. The social agency of objects, however, is meaningful only to the extent to which such objects are embedded within a social structure, where things, including software, circulate and operate with distinctive consequences for social structures.[19] Software's operations have real effects in the material world, yet software's agency always needs to be interpreted, projected, and ascribed by human agents, such as users, programmers, or designers. One may say, in this sense, that the daemon's agency is objective at a mechanical level (since its operations have real effects) but subjective at a social level (since it is always human agents who project social meanings).

Consequently, to understand daemons it is necessary to look at their material effects but also at their discursive life: in other words, at how software is perceived and represented subjectively by human users. One way to do so is by employing the concept of biography, which has two distinctive meanings. Biography describes on the one hand the course of a person's life. It constitutes an act of making sense of the life of a person in terms of the passage of time; it is, in its most literal meaning, an inscribed life. The biography of a piece of software therefore entails examining the histories of its contingent material processes of becoming, which are entangled in social processes and environments. There is on the other hand a second meaning to the word *biography*. Biography is also a literary genre, a form of contingent narrative turning historical characters, events, and things into stories to be written, told, and circulated. Even when no written text is involved, people's lives are elaborated into narratives that circulate through different channels, genres, and ways. Turned into narrative, the lives of individuals construct new meanings and signify other things. Biographies in fact embody particular representations of a person's character, temperament, personality, and skills, as well as of broader notions related to the character's profession and agency. In this second sense, unveiling the biography of daemons and other software entails examining the changing ways software has been perceived, described, and discussed. These bear the signs of the different attributes that people have projected onto them throughout their "biographical" life.[20]

Consider, then, the origins of computer daemons through both meanings of the word *biography*. Daemons were introduced as distinctive programming features when researchers working at MIT's Project MAC, led by Fernando Corbato, needed to implement autonomous routines to handle the functioning of their new time-sharing systems. Daemons were programmed in this context as pieces of software that worked in the background to seamlessly administer crucial tasks in the system without requiring a user to operate them directly. Yet in this context Corbato's decision to call them *daemons* reveals the tendency to attribute some form of humanity to them, despite the fact that they carried out rather simple and mundane tasks in the system. Recall that in ancient myths, daemons were believed to possess will, intention, and the capacity to communicate. One might contend that because Corbato had a background in physics, the concept of daemon was chosen because of the meaning of its metaphor in science, with a different connotation therefore than in Greek mythology. Yet Maxwell (whose thought experiment on the second law of thermodynamics contributed, as mentioned earlier, to the introduction of the term *daemon* in physics) lamented that the concept of the daemon resulted in humanizing agency in his thought experiment. He noted that he had not explicitly called the protagonist of his thought experiment a daemon, this characterization being made by others, while a more exact description would have been an automatic switch or a valve rather than an anthropomorphized entity.[21] Even in physics, therefore, the use of the term was viewed suspiciously, as betraying the temptation to attribute agency to imaginary devices included in thought experiments.

Corbato's naming choice, in this regard, was an act of projection through which a degree of consciousness and intelligence was implicitly attributed to daemons—understood as lines of code capable of exercising effects in the real world. The significance of this initial choice is even more evident when one examines wider trajectories in the daemon's "biography." As argued by Andrew Leonard, it is difficult to distinguish rigidly between daemons and some of the bots that later inhabited online communities, chatrooms, and other networked platforms, since the routine tasks usually attributed to daemons are sometimes taken up by chatbots with some form of conversational ability.[22] Not unlike daemons, bots are also employed to execute tasks needed to preserve the functioning of a system. These include, for instance, preventing spam in forums, posting updates automatically to social media, and contributing to moderation in chatrooms. This servility, shared by many daemons and bots, reminds us of the way voice assistants such as Amazon's Alexa are presented as docile servants of the house and its family members.

Yet the key similarity between daemons and bots is perhaps to be found not so much in what they do but in how they are seen. One of the differences between daemons and bots is that the former work "silently" in the background, while the latter are usually programmed so that they can enter into interaction with users.[23] Once embedded in platforms and environments characterized by a high degree of social exchanges, however, daemons resist this invisibility. As shown by McKelvey, for instance, the results of daemons' activities in online platforms have stimulated complex reactions. Propiracy, anticopyright groups in The Pirate Bay, for instance, antagonized daemons as well as copyright holders, making specific efforts to bring to the surface the invisible work of daemons.[24] Also in contrast to bots, daemons are not expected to engage linguistically with users' inputs. Yet, as shown by the very term chosen to describe them, daemons also invite users and developers to regard them as agents.

The biography of computer daemons shows that software, in the eyes of the beholder, takes up a social life even when no apparent interaction is involved. Like stones, computer daemons do not talk. Yet even when no explicit act of communication is involved, software attracts people's innate desire to communicate with a technological other—or conversely, the fear of the impossibility of establishing such communication.[25] Daemons, which once inhabited the middle space between gods and humans, are today suspended between the user and the mysterious agency of computer operations, whose invisible character contrasts with the all-too-real and visible effects it has on material reality.[26] Like ghosts, daemons appear unfamiliar, existing in different dimensions or worlds, but still capable of making us feel some kind of proximity to them.

THE LONG HISTORY OF DIALOGUE TREES (OR, WHY DIGITAL GAMES DON'T LIKE CHATBOTS)

Hidden routines such as daemons are capable of stimulating attributions of agency. However, the social life of AI becomes even more apparent when software permits actual interactions between computers and users. One of the applications in which this occurs more systematically is the digital game.

Despite their seemingly frivolous character (or perhaps, because of it), digital games have constituted an extraordinary arena to develop and experiment with the interactive and communicative dimensions of AI. Turing intuited this when he proposed the Imitation Game or, even earlier, when he suggested chess as a potential test bed for AI. In a lecture to the London Mathematical Society in 1947, he contended that "the machine must be allowed to have contact with human beings in order that it may adapt itself to their standards. The game of chess may perhaps be rather suitable for this purpose, as the moves of the machine's opponent will automatically provide this contact."[27] Turing was in search of something that could work as propaganda for a nascent discipline, and chess was an excellent choice in this regard because it could place computers and players at the same level in an intellectually challenging competition. But Turing's words indicate more than an interest in demonstrating the potential of AI. The development of "machine intelligence" required pathways for the computer to enter into contact with human beings and hence adapt to them, and games were the first means envisioned by Turing to create this contact.

In the following decades, digital games enhanced the so-called symbiosis between computers and humans and contributed to immerse AI systems in complex social spaces that nurtured forms of interaction between human users and computational agents.[28] While pioneering digital games such as *Spacewar!*—a space combat game developed in 1962— involved challenges between human players only, computer-controlled characters were gradually introduced in a number of games. One of the forms of interaction that became available to players was communication in natural language. To give one instance, a player of an adventure game—a genre of digital games that requires players to complete quests in a fictional world—may initiate a dialogue with a computer-controlled innkeeper in the hope of gaining useful information for completing the game quest.[29] For the history of the relationship between AI and human-machine communication, similar dialogues with fictional characters represent an extremely interesting case. Not only, in fact, do they entail conversations between users and artificial agents but also these are embedded in the fictional environment of the gameworld, with its own rules, narratives, and social structures.

Yet, interestingly, digital games rarely make use of the same kind of conversational programs that power ELIZA and other chatbots. They usually rely, instead, on a simple yet apparently very effective—at least in game scenarios—programming technique: dialogue trees. In this modality of dialogue, players do not type directly what should be said by the character they control. Instead, they are provided with a list of predetermined scripts they can choose from. The selected input activates an appropriate response for the computer-controlled character, and so on, until one of the dialogue options chosen by the player ignites a certain outcome. For instance, you, the player, are attacked if you selected lines of dialogue that provoked the interlocutor. If different options were chosen, the computer-controlled character could instead become an ally.[30]

Take *The Secret of Monkey Island*, a 1990 graphic adventure game that rose to the status of a classic in the history of digital games. In the role of fictional character Guybrush Threepwood, you have to pass three trials to fulfill your dream of becoming a pirate. The first of these trials is a duel with the Sword Master, Carla. Preparing for this challenge, you discover that sword fighting in the world of *Monkey Island* is not about dexterous movements and fencing technique. The secret, instead, is mastering the art of insult, catching your opponent off guard. During the swordplay, a fighter launches an insult such as "I once owned a dog that was smarter than you." By coming up with the right response ("He must have taught you everything you know") you gain the upper hand in the combat. An inappropriate comeback instead leads to a position of disadvantage. In order to beat the Sword Master, you need to engage in fights with several other pirates, learning new insults and fitting responses to be used in the final fight (fig. 4.1).[31]

In the *Monkey Island* saga, conversation is therefore conceived as a veritable duel between the machine and the human. This is to some extent similar to the Turing test, in which the computer program wins the Imitation Game by deceiving the human interrogator into believing it is a human. In contrast with the test, however, it is the human in *Monkey Island* who takes up the language of the computer, not the computer who imitates the language of humans. Envisaging conversation as a hierarchical structure

Figure 4.1 Sword-fighting duel in *The Secret of Monkey Island*: only by selecting the right insult will the player be able to defeat opponents. Image from https://emotionalmultimediaride.wordpress.com/2020/01/27/classic-adventuring-the-secret-of-monkey-island-originalspecial-edition-pc/ (retrieved 7 January 2020).

of discrete choices, dialogue trees in fact take up the logic of software programming: they are equivalent to decision trees, a common way to display computer algorithms made up of conditional choices. In *Monkey Island*'s "conversation game," the player beats the computer-controlled pirate by navigating software's symbolic logic: *if* the right statement is selected, *then* the player gains an advantage; *if* the player selects the right statement again, *then* the duel is won.

Dialogue trees, in other words, speak the language of computer programming, whose locution does not merely convey meaning but also executes actions that have real consequences in the material world.[32] This fits with a long-standing theory in computer science that posits that natural language, when employed to operate tasks on a computer, can be regarded as just a very high-level form of programming language.[33] Although this idea has been criticized with some reason by linguists, the equivalence between natural language and programming language applies to cases such as dialogue trees, in which the user does not need to master programming but just uses its everyday language as a tool to interact with computers.[34] In textual-based digital games, such as interactive fiction—a genre of digital games that originated as early as in the 1970s—this logic does not only concern dialogues, since language is the main tool to act on the gameworld: the player employs verbal commands and queries to act on the game environment and control fictional characters. For example, a player types GO SOUTH to have the character move in that direction in the fictional world. Something similar applies to voice assistants, with which users can carry out a simple conversation, but perhaps even more importantly, they can use language to make them perform tasks such as making a phone call, turning off the light, or accessing the Internet.

As observed by Noah Wardrip-Fruin, nevertheless, the most remarkable thing about dialogue trees is "how little illusion they present, especially when compared with systems like ELIZA."[35] Not only, in fact, dialogue trees are framed in a static dynamics of turn-taking, which reminds one more of the dynamics of table games (where players take turns to make their moves) than of real-life conversations. Furthermore, each character can only respond to a limited number of queries, with the resulting dialogues appearing at the very least rigid and, in the worst cases, overly predictable. It is surprising, and in a way fascinating, that such a rigid frame is still able to exercise a powerful effect on players—as the persistent success of dialogue trees, which continue to be widely privileged over chatbots in game design, testifies. Game scholar Jonathan Lessard, who designed a number of simple digital games experimenting with the use of chatbots, speculated that this might have to do with the fact that dialogue trees allow game designers and companies a higher degree of control of the game narrative.[36] Yet while this is likely true, dialogue trees could have not proved so resilient unless players, not only designers, appreciated this modality of play.

From this point of view, at least two additional explanations can be offered. The first one has to do with the fact that the credibility of a conversation depends on the content of communication as well as on its context. Programming ELIZA, Joseph Weizenbaum modeled conversation as a set of discrete exchanges in which context played almost no part. In contrast, dialogue trees are all about contexts: each conversational state is located at a given point in the hierarchical structure of predetermined steps, with the player's selection activating a new branch of the dialogue tree. This also means that it is impossible to go backward (although it might be possible to loop back to the main trunk of the conversation). In dialogue trees, therefore, conversation appropriates the mechanic of the player's quest: like any other action of the player, every choice is aimed at progress in the mission the player needs to accomplish.[37] Such continuity with the wider mechanism of quest-based games helps explain, at least partially, why dialogue trees are such a natural fit to genres such as adventure games or role playing. In literary fiction, after all, readers find characters compelling also due to their integration in fictional worlds and in specific scenes where they participate to the action and interact with other fictional persona.[38] The same dynamics applies to a game setting, where the appeal of a computer-controlled character derives from its being embedded in the wider mechanic of the gameplay.[39]

The second explanation has to do with the experience of playing itself. The emergence of convincing effects of personality in computer-controlled characters, even in the absence of elaborate interactions, testifies to the deep engagement computer games are able to inspire in players.[40] For instance, the computer-controlled character of the thief was the most well-known feature of *Zork*, one of the first commercial instances of the interactive fiction genre, despite the fact that its characterization was superficial to say the least: the thief interacted with players just by stealing objects and treasures or occasionally attacking them.[41] This is because the appeal of interactions with nonplaying characters in digital games depends not much on superficial similarities between interactions in the gameplay and in everyday life and more on the playful frame through which the interaction is interpreted.[42] As mentioned in chapter 3, Turkle described the Eliza effect as a form of complacency toward AI: users defused ELIZA's shortcomings by asking only those questions it was able to answer.[43] Playfulness provides, consciously or unconsciously, a strong motivation to act in this way, and therefore works as an enhancer of the effect of characterization, stimulating players to fill the gaps in the illusion spelled out by the digital game. One need only look at popular culture fans' engagements with character bots inhabiting social media, or at the teasing jokes many users attempt with AI voice assistants, to realize how this happens even outside of game environments.[44]

Reportedly, the main satisfaction of Will Crowther, the creator in the 1970s of the first textual-based interactive fiction game, *Adventure*, was to discover that his program fooled people into thinking that it was intelligent enough to understand and speak English. As a colleague recounted, "Will was very proud—or more accurately amused—of how well he could fool people into thinking that there was some very complex AI behind the game."[45] The similarity of this comment with Weizenbaum's reflections about ELIZA are noteworthy. Like Weizenbaum, Crowther noticed that users tended to overstate the proficiency of relatively simple computing systems if they were processing and generating natural language.[46] This pointed to something akin to the deception Weizenbaum noted in users of his ELIZA: a propensity to regard even unsophisticated manipulation of language as evidence of "intelligence" on the part of computers.

MICROSOFT BOB, OR THE UNFORTUNATE RISE OF SOCIAL INTERFACES

In terms of banal deception, human-computer interaction can be conceptualized as the gradual development of tools to "reach out" toward human users, adapting computer interfaces to the perceptive and cognitive abilities of mankind.[47] Although this process has been anything but a linear evolution, interface designers have made constant efforts to incorporate knowledge of how humans receive, process, and memorize information in order to foster more effective interactions with computers. The development of networked systems, therefore, worked as an incentive for designers to consider that every interaction with a computer was a social activity that should be understood against the background of social expectations and understandings. Understandings of the social world were increasingly included as frameworks for the development of interfaces, in the hope of exploiting familiarity with social situations in the everyday world.[48]

In this context, the development of AI-based software that processes natural language presented designers with the opportunity to develop interfaces that literally entered into conversation with users.[49] Alexa, Siri, and Google Assistant are the latest instances of this approach, but the history of "social interfaces" or "user-interface agents" is much longer.[50] In the mid-1990s Microsoft, the leading company in the computer sector at the time, made significant efforts to innovate in this area. Their first social interface project was Microsoft Bob, an animated cartoon-character-based user interface that was introduced in March 1995 and whose production was discontinued just one year later.

Microsoft Bob was, by all definitions, a failure. In computer science circles, it has acquired a status similar to the infamous film director Ed Wood among cinephiles: a standard example of something that has gone badly wrong. As one tech journalist put it, "What's the most efficient way to deride a technology product as a stinker and/or a flop? Easy: Compare it to Microsoft Bob."[51] Announced as a revolution in user-friendliness and as the embodiment of the new paradigm of social interfaces, Bob was sharply criticized by critics and rejected by users, who found it condescending and unhelpful.[52] Yet the system's failure does not make it a less valuable case to explore. On the contrary, as historians of science and technology have shown, failed technologies are often ideal entry points for examining critically decisive turns and trajectories of technological evolution.[53] In the history of personal computing, Microsoft Bob is a quite interesting case: drawing from insights in the social sciences, a major company developed a new product that featured an imaginative vision of the future of interface design. Its spectacular failure provided an opportunity for practitioners in the field to critically consider both the benefits and the challenges of social interfaces. The success of other social interfaces in the following two decades, moreover, demonstrates that Microsoft's project was not a dead end but an ill-fated project reliant on some misconceived ideas as well as on fertile ground.[54]

Introduced at a time when the Microsoft company was the undiscussed leader of the software industry, Bob was promoted with much fanfare as a breakthrough that would revolutionize home computing. Its logo, a smiling face with heavy glasses, featured in marketing items that included sports watches, baseball caps, and T-shirts.[55] Bob was an integrated collection of home computing applications whose interface was based on the visual metaphor of a house with rooms and contents, through which users could access applications and functions. It included eight interconnected programs, including email, a calendar, a checkbook, an address book, and a quiz game. Designed with a cartoon graphic, it was targeted to children and nonexpert users, in an attempt to enlarge computers' user base.

Microsoft's press releases stressed that Bob didn't look like any traditional software because users were to interact with Bob "socially." This was made possible through "personal guides" that responded to inputs through dialogue balloons and provided users with "active and intelligent help" as well as "expert information." The guides were presented in a virtual environment that simulated domestic spaces and could be personalized by users, "decorating" the rooms and even changing the scenery outside the rooms' windows to suit their taste.[56] Users could choose from a list of twelve creatures, the default being a dog named Rover, and other choices, including a French-sounding cat, a rabbit, a turtle, and a sullen rat.[57] The guides sat at the corner of the screen, providing instruction and performing occasional gimmicks while users were using the software (fig. 4.2). One of the features Microsoft boasted about was the degree to which the different "personal guides" were characterized by specific personalities. When the user was given the choice of selecting a personal guide, each was described with attributes depicting its personality, for example, "outgoing," "friendly," or "enthusiastic."[58] Although the experience with all guides was in truth quite similar, some of the lines of dialogue were specific to each one's personality.

Like the metaphor of the domestic space, the name "Bob" was purportedly chosen to sound "familiar, approachable and friendly."[59] The reaction of expert reviewers and journalists, however, was not friendly at all. Bob was described as "an embarrassment," and criticized for evident flaws in its design.[60] In the virtual rooms, for instance, there were objects giving access to applications or to certain functions. However, other objects were useless

Figure 4.2 Screenshot of Microsoft Bob, with the guide "Rover the dog" (lower right). Image from https://theuijunkie.com/microsoft-bob/, retrieved 24 November 2019.

and, when clicked on, delivered a message explaining that they had no particular function. This was hardly an example of seamless design.[61] Another common criticism revolved around the fact that the interface offered "indirect management," meaning that it not only gave access to functions but also performed operations on behalf of the user. This was seen as potentially problematic especially because the applications offered in Microsoft Bob included home finance management, which raised the question of who would be accountable for eventual mistakes.[62]

Arguably, the main reason for Bob's failure, however, was that the interface struck users as intrusive. Despite its having been explicitly programmed to facilitate social behavior, Bob did not follow important social rules and conventions concerning things such as conversation turns and personal space. The "personal guides" frequently interrupted navigation, with help messages bubbling up unrequested. As one reviewer put it, "Bob's programs eliminate many standard functions, bombard users with chatty messages and do not necessarily make it easier to do things."[63] In addition, the didactic tone of the program and its personal guides were irritating to many. Shortly after Bob's release, a programmer named George Campbell created a parody of Bob with its helpers offering "little bits of drawling advice—more typically about life than computers—such as: 'Listen to the old folks; y'all might get to live as long." Or: 'A thirsty man oughtn't to sniff his glass.'"[64]

Yet for all the derision it attracted later, Bob did not look ridiculous at the time. Its concept originated in what has been probably the most authoritative social sciences approach to human-computer interaction developed over the last three decades, the Computers Are Social Actors paradigm. Clifford Nass and Byron Reeves, whose studies built the foundations of this approach, consulted for Microsoft and participated in events connected to Bob's launch.[65] Famously, Nass and Reeves argue that people applied to computers and other media social rules and expectations similar to those they employed when interacting with people in their everyday lives.[66] Translated into the complexity of interface design, their research indicated that each component in interface design, up to the way a message was phrased, conveyed social meaning. The possibility of anticipating and directing such meaning could be exploited by designers to build more effective and functional interactions with computers.[67]

Nass gave Bob passionate endorsement in press releases and interviews, underlining that Bob was a natural development of the findings developed through Nass's and Reeves's own research: "our research shows that people deal with their computers on a social level, regardless of the program type or user experience. People know that they're working with a machine, yet we see them unconsciously being polite to the computer, applying social biases, and in many other ways treating the computer as if it were a person. Microsoft Bob makes this implicit interaction explicit and thereby lets people do what humans do best, which is to act socially."[68] Reeves was later also quoted in a press release as stating: "the question for Microsoft was how to make a computing product easier to use and fun. Cliff and I gave a talk in December 1992 and said that they should make it social and natural. We said that people are good at having social relations—talking with each other and interpreting cues such as facial expressions. They are also good at dealing with a natural environment such as the movement of objects and people in rooms, so if an interface can interact with the user to take advantage of these human talents, then you might not need a manual."[69]

Looking at Microsoft Bob alongside Nass's and Reeves's theories makes a quite interesting exploration of the challenge of bringing together theory with practice. On the one side are the spectacular achievements of Nass's and Reeves's paradigm, which has informed a generation of research as well as practical work in human-computer interaction; on the other side the spectacular failure of an interface, Microsoft Bob, which was inspired by this very same paradigm. How can one make sense of the relationship between the two?

Perhaps the key to Microsoft's inability to convert Nass's and Reeves's insights into a successful product is in Nass's statement that Microsoft Bob represented the ambition to take the *implicit* social character of interactions between humans and computers and make it *explicit*. Yet in Nass's and Reeves's research, as presented in their groundbreaking book *The Media Equation,* published shortly after the launch of Microsoft Bob, the social character of interactions with computers is implied and unspoken, rather than acknowledged or even appreciated by the user.[70] In his later work, Nass has further refined the Computers Are Social Actors paradigm through the notion of "mindless behavior." Through the notion of mindlessness, drawn from the cognitive sciences, he and his collaborator Youngme Moon have emphasized that users treat computers as social actors despite the fact that they know all too well that computers aren't humans. Social conventions derived from human-human interaction are thus applied mindlessly, "ignoring the cues that reveal the essential asocial nature of a computer."[71]

In Nass's and Reeves's theoretical work, therefore, computers' sociality is not intended to be explicit and constrictive as was the case in Microsoft Bob. The intrusiveness of Bob, in this sense, might have disrupted the mindless nature of imaginative social relationships between humans and machines. To remain more strictly within the boundaries of the history of human-computer interaction, one could say that Bob's "explicit" sociality was a blatant contravention of the logics of transparent computing.[72] At Microsoft, the Bob project had originated from the same team who had previously developed another software package, Microsoft Publisher, which is still in use today. Introduced in 1991, Publisher was the first Microsoft application to use wizards that conducted users through complicated tasks step by step. Microsoft Publisher's designers regarded the social interface of Microsoft Bob as a logical next step toward the goal of making software more approachable to newbies at a time when most households still didn't have a personal computer.[73] Yet extending the didactic approach of Publisher's wizards to an entire set of applications was problematic. While wizards intervene when certain operations are required, Bob's personal guides constantly affirmed their presence, lacking the capacity to disappear quietly in the background. The guides asked users to take them seriously as social agents, refusing to accommodate the subtle play between hiding and revealing, appearing and disappearing, that other computer interfaces navigate.[74] If an interface always builds to some extent an illusion, the way Nass's and Reeves's ideas were implemented turned Bob into an interface where the illusion was never concealed but always made manifest to the user.[75]

FOR A GENEALOGY OF ALEXA

Searching on YouTube for videos about Alexa, one finds a large number of homemade videos of users engaging in conversations with "her." A veritable subgenre among them are videos that play jokingly with weird or comic replies returned by Alexa to specific inputs. Some, for instance, hint at the famous *Star Wars* scene in which Darth Vader tells young hero Luke Skywalker "I am your father." In the YouTube clips Alexa responds to users with the same line, certainly scripted by Amazon developers: "Noooo! That's not true, that's impossible."[76]

In spite of Alexa's resistance to accepting its users' paternity, Alexa, Siri, Google Assistant, and other contemporary voice assistants have not one but many "fathers" and "mothers." This chapter has tracked three crucial threads that are part of this story: daemons, digital games, and social interfaces. Histories of AI are usually more conservative in defining the precedents of today's AI communicative systems, such as voice assistants. In technical terms, their software is the result of a long-standing evolution in at least two main areas within AI: natural language processing and automated speech recognition and production. On the one hand natural language processing develops programs that analyze and process human language, from chatbots to language translation, from the interpretation of speech to the generation of text.[77] On the other hand automated speech recognition and production develops computer applications that process spoken language so that it can be analyzed and reproduced by computers. The genealogy of voice assistants is, however, much more complex. Since, as Lucy Suchman has taught us, every form of human-computer interaction is socially situated, this genealogy also involves practical and mundane software applications through which AI has developed into technical, social, and cultural objects.[78]

Each of the three cases examined in this chapter provide alternative entry points to interrogating the subtle and ambivalent mechanisms through which social agency is projected onto AI. Obscure and hidden by definition, computer daemons testify to the dynamics by which autonomous action leads to attributions of agency and humanity. Notwithstanding the routine character of the operations daemons perform, their name and the "biography" of their representations help locate computing machinery in an ambiguous zone between the alive and the inanimate.

The role of digital games in the evolution of AI has often been overlooked. Yet the playful character of games has permitted safe and imaginative exploration of new forms of interactions and engagements with computers, including dialogues with AI agents and computer-controlled characters. As in digital games, modern interactions with voice assistants elicit and stimulate playful engagements through which the boundaries between the human and the machine are explored safely and without risk. Examining the uses of dialogue trees helps one appreciate the role of playfulness in interacting with communicative AI, where part of the pleasure comes from falling into the illusion of entertaining a credible conversation with non-human agents.

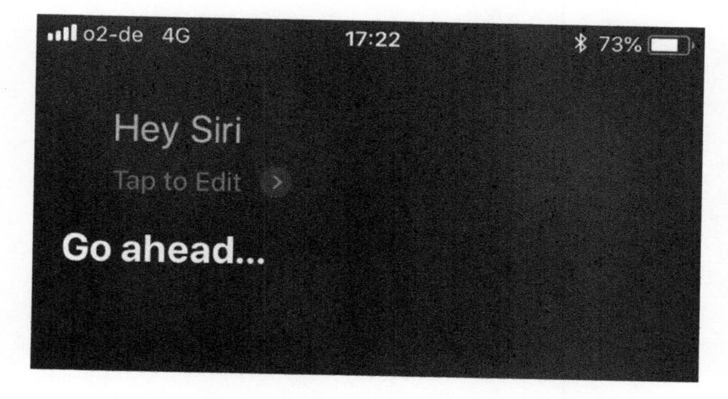

Figure 4.3 Author's conversation with Siri, 12 December 2019.

Finally, the case of a failed experiment with social interfaces, Microsoft Bob, provides ground for reflection on what makes AI-based interfaces a nonthreatening presence in our domestic environments and everyday lives. Microsoft's attempt to make explicit the sociality of our interactions with computers backfired, as Bob's personal guides were perceived as intrusive and annoying. This contrasts with the reserved behavior of today's voice interfaces, which remain silently and (apparently) inactive in the background until the wake word calls them to life (fig. 4.3):

CHAPTER 5
How to Create a Bot
Programming Deception at the Loebner Prize Competition

American entrepreneur Hugh Loebner was by all accounts a controversial character. The holder of six patents and the owner of Crown Industries, a manufacturer of theatre equipment, he made headlines with his vocal opinions in favor of prostitution, as well as the role he tailored for himself in the search for AI. In the late 1980s, Loebner became excited by the idea that the Turing test should be made into an actual contest, and he eventually decided to fund such an enterprise himself. He partnered with Robert Epstein, a cognitive psychologist and director of the Cambridge Center for Behavioral Studies, to bring his dream into reality. The first edition of the contest, called the Loebner Prize, was organized in 1991. The contest has been conducted every year since then in several locations around the world, from the United Kingdom to Australia, from Loebner's own apartment in New York City to the facilities where Alan Turing worked during World War II to crack ciphered Nazi messages at Bletchley Park.

In line with Loebner the public figure, the Loebner Prize has also been highly contentious. While its supporters have described it enthusiastically as a part of the "quest for the thinking machine," many have sharply criticized the contest, arguing that it does not provide an appropriate setting for evaluating the achievements of cutting-edge AI technologies.[1] Critics point to the fact that the contest only measures language proficiency in a conversation setting, which is just one niche in the wide field of

Deceitful Media. Simone Natale, Oxford University Press (2021). © Oxford University Press.
DOI: 10.1093/oso/9780190080365.003.0006

AI applications, and that even in this specific area the contest does not reflect the most advanced technologies.[2] They note that the contest does not improve understanding of the discipline but on the contrary creates hype and supports misleading views of what AI actually means.[3] Persuasively, they also contend that what makes computer programs successful at the Loebner Prize is largely their capacity to draw on human fallibility, finding tricks and shortcuts to achieve the desired outcome: fooling people into thinking that a bot is human.[4]

While I agree with such criticisms, I also believe that paradoxically, the very reasons why the Loebner Prize has little relevance as an assessment of technical proficiency make it an extremely productive case for understanding AI from a different perspective. As I have shown, when AI systems enter into interactions with humans, they cannot be separated from the social spaces they share with their interlocutors. Humans need to be included in the equation: communicative AI can only be understood by taking into account issues such as how people perceive and react to the behavior of AI. In this regard, the Loebner Prize provides an extraordinary setting where humans' liability for deception and its implications for AI are put to the test and reflected on. This chapter looks at the history of this competition to argue that it functioned as a proving ground for AI's ability to deceive humans, as well as a form of spectacle highlighting the potential but also the contradictions of computing technologies. Although the prize has not offered any definitive or even scientifically valid assessment of AI, the judges' interactions with humans and programs during the nearly thirty years of the competition represent an invaluable archive. They illuminate how conversational bots can be programmed to trick humans, and how humans respond to the programmers' strategies for deception.

ON WITH THE SHOW: PERFORMING THE TURING TEST

It wasn't originally meant to be for real. In 1950, when Turing proposed his test, he conceived it as a thought experiment, a provocation that would make people realize the potential of computing. In the following decades, his proposal stimulated lively theoretical discussions in areas such as psychology, computer science, and the philosophy of mind, but little efforts were made to bring it into actual performance. Why, at the beginning of the 1990s, did someone decide to organize an actual public contest based on the Turing test?

It is tempting to answer this question by pointing to technological progress. This would mean to assume that in the early 1990s AI had finally reached such a level of sophistication and proficiency that it was worth putting it to the Turing test. This hypothesis, however, is only partially convincing. Although new generations of natural language processing programs had brought to fruit the theory of generative grammar, conversational agents such as chatbots had not evolved dramatically since the times of ELIZA and PARRY.[5] In fact, even the organizers of the first Loebner Prize competition did not expect computers to come anywhere close to passing the Turing test.[6] Further, their actual performance hardly demonstrated clear technical advances. As several commentators noted, contestants went on to use many of the techniques and tricks pioneered by the ELIZA program, developed almost thirty years earlier.[7]

What led to the organization of the Loebner Prize competition were probably not technical but mainly cultural factors.[8] When Epstein and Loebner teamed up to convert Turing's thought experiment into an actual contest, the growth of personal computing and the Internet had created ideal conditions for carrying out such an endeavor. People were now accustomed to computing machines and digital technologies, and even chatbots and AI characters were not unknown to Internet users and digital game players. After roughly two decades of disillusion and the "AI Winter," fertile ground was laid for renewing the myth of thinking machines. The climate of enthusiasm for computers involved the emergence of a powerful narrative around the idea of the "digital revolution," with the help of enthusiastic promoters, such as Nicholas Negroponte, founder of the MIT Media Lab in 1985, and amplifiers, like the popular magazine *Wired*, which debuted in 1993.[9]

As part of this climate, computers moved to the center stage not just of the Loebner Prize but of several human-versus-machine competitions. From chess to popular quiz shows, the 1990s saw computer programs battling against humans in public contests that attracted the attention of the public and the press.[10] It was not a coincidence, in this sense, that Epstein was a skilled science popularizer and Loebner the owner of a firm that specialized in theatre equipment. The Loebner Prize was first and foremost a matter of public communication and performance; as Margaret Boden put it, "more publicity than science."[11] This was not completely in contradiction with the intentions of Turing himself: as noted in Chapter 1, Turing had in fact intended his original proposal of the test to be a form of "propaganda" to raise the status of the new digital computers.[12] It is entirely appropriate, in this sense, that the Loebner Prize was aimed primarily at appealing to the public and only secondarily as a contribution to science.

The first competition for the Loebner Prize was hosted on November 8, 1991, by the Computer Museum in Boston. Oliver Strimpel, the museum's executive director, explained to the *New York Times* that the event met one of the key goals of the museum: "to answer the 'so what?'—to help explain the impact of computing on modern society." He also explicitly acknowledged the continuity between the Loebner Prize and other AI contests, noting that the museum hoped to organize another "human vs machine showdown": a match between the world's chess master and the most advanced chess-playing computer.[13]

Due to the vagueness of Turing's proposal, the committee that prepared the event had to decide how to interpret and implement his test. They agreed that judges would rank each participant (either a computer or a human) according to how humanlike the exchanges had been. If the median rank of a computer equaled or exceeded the median rank of a human participant, that computer had passed this variant of the Turing test. In addition, judges would specify whether they believed each terminal to be controlled by a human or by a computer. The judges were to be selected among the general public, and so were the "confederates," that is, the humans who would engage in blind conversations with judges alongside the computer programs.

Another important rule concerned the conversation topic: the scope of the test was limited to a specific theme, to be chosen by each contestant. Epstein admitted that this rule, which changed in 1995 when discussions became unrestricted, was meant "to protect the computers" that were "just too inept at this point to fool anyone for very long."[14] Other regulations were also conceived to the computers' advantage: judges, for instance, were solicited through a newspaper employment ad and screened for having little or no knowledge of computer science, which heightened the chances of the chatbots' programmers.[15]

For a contest aimed at measuring AI progress, it seems odd that specific regulations were aimed at facilitating the programmers' work, jeopardizing the impartiality of the proceedings. Yet this choice is reasonable if one concedes that the Loebner Prize competition is not so much scientific research as a form of spectacle. Just as the organizers of a sporting event need athletes to be in their best shape to compete, the prize's committee need computers to be strong contestants in their game—hence the decision to easen their task.[16] The need to enhance the spectacularism of the competition also explains the decision to award a bronze medal and a cash prize every year to the computer program that, according to the judges, has demonstrated the "most human" conversational behavior—even if the program has failed to pass the Turing test. The bronze medal was intended not only as a motivation for participants but also as a further concession to the spectacular character of the event: the presence of a winner, in fact, fits the familiar frame through which public competitions are presented to journalists and audiences.[17]

There is a long tradition of exhibiting scientific and technological wonders to the public. In the history of computing, public duels featuring machines and humans, such as chess competitions between human champions and computer programs, have often transformed the ordinary and invisible operations of machines into a sensational spectacle.[18] The automaton chess player that stunned observers in the eighteenth and nineteenth century also exploited the appeal of machines battling against humans in what Mark Sussman called "the faith-inducing dramaturgy of technology."[19] Faithful to this tradition, the Loebner Prize competition was organized as a spectacle and hosted in a large auditorium before a live audience. A moderator wandered around the auditorium with a cordless microphone, interviewing some of the contestants and commenting on their competing chatbots, while audiences could follow the conversations in screen terminals. The spectacular frame of the event reverberates in the reports describing the first competition as "great fun" and an "extravaganza."[20]

Substantial attention was given to promoting the prize to the press. The date was anticipated by a robust press campaign, with articles published in some of the most widely read newspapers on both sides of the Atlantic.[21] The need to make the contest easily understandable to journalists was mentioned among the criteria orienting the committee in deciding on the regulations.[22] As reported by one of the referees, Harvard University computer scientist Stuart Shieber, "the time that each judge had to converse with each agent was shortened from approximately fifteen minutes to approximately seven in order to accommodate the press's deadlines."[23] After the event, three articles appeared in the *New York Times* alone, including a front-page article the day after the contest, and Epstein boasted that the winner had received the equivalent of a million dollars in free advertising thanks to the press coverage of the event.[24]

The circulation and dissemination of narratives about technology in the public sphere tend to follow recurring patterns, relying on repetition that shapes these narratives' very capacity to become popular and widespread.[25] This process has been described in Latourian terms as a "pasteurization" of the narrative, by which elements that do not fit into the dominant narrative are eliminated so as to privilege a more coherent and stable narrative, exactly like germs being eliminated in food pasteurization processes.[26] In the press coverage of early Loebner Prize competitions, it is easy to identify the recurring script through which the event was narrated. The narrative posited that even if no computer had been able to pass the Turing test, encouraging successes could be observed that suggested a potential future in which computers would pass the test and become indistinguishable from humans.[27] This script suited the overarching narrative of the AI myth, which since the early years had pointed to promising if partial achievements of AI to envision a future where human-level AI would be reached, changing the very definition of humanity and the social world.[28] The opposing narrative, according to which AI was a fraud, also emerged in more critical reports of the prize. Thus the coverage of the prize drew from and perpetuated the long-standing dualism between enthusiasm and criticism that has characterized the entire history of AI.[29]

By contrast, more ambiguous implications of the prize's procedures that did not fit the classic AI controversy were rarely, if ever, discussed in public reports. The emphasis was on computers: what they were able to do and to what extent they could be considered "intelligent." Much less attention was given to the humans who participated in the process. This is unfortunate because, as shown in chapter 1, the Turing test is as much about humans as it is about computers. Yet only a few newspaper pieces mentioned that not only computers were exchanged for humans at the contest: humans were often exchanged for computers, too. For instance, in the first competition, the confederate who impressed the judges as "the most human of all [human] contestants" was still deemed to be a computer by as many as two judges.[30] This pattern continued in the following years, persuading the prize's committee to assign a nominal prize every year to the confederate judged "the most human human."[31]

If the fact that the judges could struggle to recognize the humanity of confederates seems counterintuitive, one should remember that not all instances of human communication conform to widespread ideas of what counts as "human." Think, for instance, of highly formalized interactions such as exchanges with a phone operator, or the standardized movements of telegraphists, whose actions were later automated through mechanization.[32] Even in everyday life, it is not uncommon to describe a person who repeats the same sentences or does not convey emotion through language as acting or talking "like a machine." To give a recent example, one of the moments when it became evident that the 2017 election campaign was not going well for British prime minister Theresa May was when newspapers and users on social media started to refer to her as the "Maybot." By likening her to a bot, journalists and the public sanctioned her dull and repetitive communication style and the lack of spontaneity in her public appearances.[33]

The sense that some communications are "mechanical" is amplified in contexts where exchanges are limited or highly formalized. Most exchanges between customers and a cashier in a big supermarket are repetitive, and that is why this job has been mechanized faster and more effectively than, say, psychotherapy sessions. Yet the conversation between a cashier and customer sometimes goes out of the usual register—for instance, if some jokes are exchanged or if one shares one's personal experiences with a product. These are the kinds of conversations that are perceived as more "human"—thus, the ones that both a computer and a confederate should aim for in order to convince judges that they are flesh and blood.[34]

One of the consequences of this is that within the strict boundaries of the Loebner Prize competition, people and computers are actually interchangeable. But this is not because computers "are" or "behave" like humans. It is, more subtly, because in the narrow situation of a seven-minute written conversation via computer screen, playing the Imitation Game competently is the key to passing as human.

Brian Christian, who participated as a confederate in the 2009 Loebner Prize competition and wrote a thoughtful book on the experience, was told by the organizers to be "just himself." Yet, in truth, confederates, to strike judges as humans, should not just be themselves: they need to opt for conversation patterns that will strike judges as more "human." Christian, therefore, decided to ignore the organizers' advice. In order to establish an effective strategy, he asked himself which dynamics might have informed the judges' decisions. How does a person appear "human" in the setting of the Turing test? Which strategies should be adopted by a confederate or coded into a computer to maximize the chances of passing for human? In order to answer such questions, Christian not only studied transcripts of past Loebner Prize competitions but also reflected on which behaviors are sanctioned as "mechanical" by people and, conversely, which behaviors are perceived as more authentic and "human." He went on to win the 2009 award as the "most human human," that is, the confederate who had been considered most credible by the judges.[35] With the appropriate preparation, he had become the best player of the Imitation Game.

Judges for the Loebner Prize also came up with some clever techniques for performing their roles more effectively. Some, for instance, intentionally mumbled and misspelled words or asked the same questions several times to see if the conversant could cope with repetition, as humans regularly do.[36] Such tricks, however, would only work until a clever programmer figured them out and added appropriate lines of code to counteract them. Not unlike the all-to-human confederates and judges, in fact, programmers developed their own strategies in the attempt to make their machines the most skilled contestants for the prize.

Commenting on the outcome of the first Loebner Prize competition in 1991, Epstein was not completely convinced by the winner, Joseph Weintraub, and his chatbot PC Therapist. The program, despite falling short of passing the Turing test, had clearly outshone all competitors. The problem was, however, the way it reached this result. "Unfortunately," Epstein pointed out, "it may have won for the wrong reasons."[37] Its secret weapon was the choice of the conversation topic "Whimsical Conversation." Coherently with this script, PC Therapist simulated a very particular kind of person: the jester. This proved to be a brilliant move, and five out of the ten judges ranked it as a human, despite the fact that the program exhibited quite erratic behavior. Since even impenetrable and irrelevant responses could be expected from a jester, this appeared totally logical to some judges. As Stuart Shieber commented, "Weintraub's strategy was an artful dodge of the competition rules. He had found a loophole and exploited it elegantly. I for one believe that, in so doing, he heartily deserved to win."[38]

Weintraub was not the only trickster. Chatbot developers, in fact, understood from the start that in order to pass the Turing test a computer didn't need to be "intelligent": it only needed to fake intelligence. As noted by Turkle, most efforts to develop a computer program that have passed the Turing test "have focused on writing a program that uses 'tricks' to appear human rather than on trying to model human intelligence."[39] Developers, for instance, realized that a program was more credible if it imitated not only humans' skills and abilities but also their shortcomings. A good example is the time a computer should take to respond to a user's input. Due to the calculating power of modern machines, a state-of-the-art chatbot can reply in a fraction of second. In theory, answering queries so rapidly should be regarded as evidence of skillfulness and thus "intelligence." However, because such speed is impossible for humans, chatbots are often programmed to take more time than they need for their answers. Programmers have therefore calculated how much time humans take on average to spell words on a keyboard and have programmed their chatbots accordingly.[40]

Another example is spelling mistakes. Since people often go wrong, these are also perceived as evidence of humanity. A computer contestant may thus be more credible if it makes an occasional typo or shows some inconsistency in writing.[41] This is ironic if one considers that the Loebner Prize competition promises to compensate the most skillful programs. In fact, in preparation for the 1991 competition the prize committee was reportedly "obsessed for months" over the question of whether or not they should allow entrants to simulate human typing foibles, and whether messages should be sent in a burst or in character-by-character mode. Their decision was finally to leave all possibilities open to contestants, so that this variable could become an integral part of the "tricks" exploited by programmers.[42]

Common tricks included admitting ignorance, inquiring "Why do you ask that?" to deflect from a topic of conversation, or using humor to make the program appear more genuine.[43] Some, like the team who won the 1997 competition with the chatbot CONVERSE, figured out that commenting on recent news would be regarded by judges as a positive indication of authenticity.[44] Many strategies programmers elaborated at the competition were based on accumulations of knowledge about what informed the judges' assessment. Contestant Jason Hutchens, for instance, explained in a report titled "How to Pass the Turing Test by Cheating" that he had examined the logs of past competitions to identify a range of likely questions the judges would ask. He realized for example that a credible chatbot should never repeat itself, as this had been the single biggest giveaway in previous competitions.[45]

The pervasiveness of tricks adopted by programmers to win the Loebner Prize is easily explained by the fact that the contest does not simply allow deception but is organized around it. Every interaction is shaped by the objective of fooling someone: computer programs are created to make judges believe they are human, judges select their questions in the attempt to debunk them, confederates use their own strategies to make themselves recognizable as humans. In this sense, to return to a comparison made in chapter 1, the Loebner Prize competition looks like a spiritualist séance where a fraudulent medium uses trickery to deceive sitters, while skeptics carry out their own tricks and ploys to expose the medium's deceit.[46] This setup contrasts to the conditions of scientific research: as Hugo Münsterberg, a psychologist of perception and early film theorist, pointed out, "if there were a professor of science who, working with his students, should have to be afraid of their making practical jokes or playing tricks on him, he would be entirely lost."[47] But this setup also differs from most forms of human-computer interaction. While the judges of the Loebner Prize expect that their conversation partners are computers built to simulate humans, in many platforms on the web the default expectation is that everybody is a human.[48]

This does not mean, however, that forms of deception similar to the ones observed at the Loebner Prize contest are not relevant to other contexts of human-machine communication. The history of computing is full of examples of chatbots that, like the winner of the inaugural Loebner Prize, were programmed to exhibit unpredictable or even pathological personalities. A chatbot developed in 1987 by Mark Humphrys, a student at University College Dublin, succeeded, for instance, "due to profanity, relentless aggression, prurient queries about the user," and contentions that users were liars when they responded. Humphrys's chatbot was created as a version of ELIZA but, contrary to Weizenbaum's creation, was not designed to be sympathetic but "to have an unpredictable (and slightly scary) mood."[49] For instance, in response to any innocuous statement beginning with "You are . . . (y)," the chatbot responded with one of these lines:

(i am not (y) you insulting person)
(yes i was (y) once)
(ok so im (y) so what is it a crime)
(i know i am (y) dont rub it in)
(i am glad i am (y))
(sing if youre glad to be (y) sing if youre happy that way hey)
(so you think i am (y) well i honestly could not care less)[50]

When Humphrys named the program MGonz and put it online on the preweb Internet in 1989, an element of surprise was added: unlike the Loebner Prize scenario, most users did not expect that they could be chatting with a computer agent. This made it more likely for users to believe that they were talking to a human.

In this sense, the Loebner Prize is a reminder that computer-mediated communication is a space essentially open to deception. One of the characteristics of online environments is the complex play of identity that avatars, anonymity, and pseudonyms allow. As argued by Turkle, "although some people think that representing oneself as other than one is always a deception, many people turn to online life with the intention of playing it in precisely this way."[51] MGonz's experience in the preweb era also foregrounded the more recent upsurge in online abuse and trolling. Journalist Ian Leslie recently referred to MGonz when he proposed that Donald Trump might be the first chatbot president, as "some of the most effective chatbots mask their limited understanding with pointless, context-free aggression."[52]

The points of continuity between the Loebner Prize competition and online spaces resonate in an ironic episode involving Epstein. He was himself the victim of a curious online scam when he carried on an email correspondence for months with what he thought was a Russian woman. It turned out to be a bot that employed tricks similar to those of Loebner Prize contestants, hiding its imperfections behind the identity of a foreigner who spoke English as a second language.[53] As stressed by Florian Muhle, "the most successful conversational computer programs these days often fool people into thinking they are human by setting expectations low, in this case by posing as someone who writes English poorly."[54] Today, some of that strategies that were conceived and developed by Loebner Prize competitors are mobilized by social media bots pretending to be human. In this sense the prize, promoted as a contest for the development of intelligent machines, has turned out to be more of a test bed for the gullibility of humans confronted with talking machines.

In fact, one of the common criticisms of the prize is that, as a consequence of the reliance on tricks, the programs in the competition have not achieved significant advances on a technical level and thus cannot be considered cutting-edge AI.[55] Yet the relatively low level of technological innovation makes the repeated success of many programs in fooling judges all the more striking. How could very simple systems such as PC Therapist and CONVERSE be so successful? Part of the answer lies in the circumstance, well known to confidence men, psychic mediums, and entertainers à la P. T. Barnum, that people are easily fooled.[56] As Epstein admitted, "the contest tells us as much, or perhaps even more, about our failings as judges as it does about the failings of computers."[57]

Yet being content with the simple assertion that people are gullible, without exploring it in its full complexity and in the specific context of human-machine communication, would mean ignoring what the Loebner Prize competition can really tell us. In the search for more nuanced answers, one may consider Suchman's discussion of the "habitability" problem with respect to language, that is, "the tendency of human users to assume that a computer system has sophisticated linguistic abilities after it has displayed elementary ones."[58] This might have to do with the experience, constantly confirmed in everyday exchanges with our own species, that those who master language have indeed such abilities. Until very recently, humans' only precedent in interacting with entities that use language has been almost exclusively with other humans. This helps explain why, as research in human-machine communication has demonstrated, users tend to adopt automatically social behaviors with machines that communicate in natural language.[59] Something similar also happened with the introduction of communication technologies that recorded and electrically transmitted messages in natural language. Although mediation increases the separation between sender and receiver, people easily transferred frames of interpretation that were customary in person-to-person exchanges, including empathy and affect.[60]

Art historian Ernst Gombrich—whose *Art and Illusion* is, though unbeknownst to most media theorists, a masterpiece of media theory—presents an interesting way of looking at this issue when he argues that "the illusion of art presupposes recognition." He uses the example of a drawing of the moon. A viewer's capacity to recognize this moon, he notes, has little to do with how the moon appears. It has much more to do with the fact that the viewer knows what the drawing of a moon looks like, and thus makes the right guess—instead of thinking, for instance, that the drawing portrays a cheese or a piece of fruit. Our visual habits orient recognition much more than any correspondence between the drawing and the "natural" appearance of the moon.[61] Looking at the problem of AI from this point of view, the use of language can be seen as facilitating an effect of recognition, orienting one's guess that one is communicating with an intelligent actor.

In sociological thought, Pierre Bourdieu's notion of habitus describes the set of habits and dispositions through which individuals perceive and react to the social world. Crucially, habitus allows individuals to understand new situations based on previous experiences. In the context of human-computer interactions through natural language, the habitus of previous linguistic interactions informs new experiences of exchanges with computers in written or spoken language.[62] Research, however, also shows that the initial attribution of sociality might be dispelled in subsequent interactions if the machine does not live up to expectations. As Timothy Bickmore and Rosalind Pickard note, "while it is easy to get users to readily engage an agent in social dialogue, it is an extremely challenging task to get the agent to maintain the illusion of human-like behavior over time."[63] This may suggest that the act of communication, undertaken in natural language, helps ignite the habitus ingrained in the user by previous interactions with humans. The strength of this "recognition effect" (to paraphrase Gombrich) subsequently dissolves if the AI is not able to engage with the linguistic and social conventions that are likewise embedded in individuals' habitus through socialization.

It is interesting to see how this dynamic plays out in the context of the Loebner Prize competition. Computer programs are here ranked for their capacity to maintain for a given time the recognition effect posed by their use of natural language. Their capacity to achieve this is informed by a combination of the program's internal functioning, the judge's habitus, and the Loebner Prize competition's rules of play. I mentioned earlier that some of the interpretations of the Turing test by the prize's committee were meant to give an advantage to programmers. However, one might even go beyond this statement to argue that the Turing test itself largely responds to the goal of limiting the ways in which the recognition effect might be challenged. Elements such as the short time available for the conversation and the fact that communication is carried out through the mediation of a text-only interface are meant to restrict the scope of the experience, thereby easing the computers' task. All these boundaries to the communication experience reduce the possibility that the initial recognition effect, and the illusion of humanity, will erode.[64]

Many of the programmers' tricks are also directed at limiting the scope of conversations so that potential hints that the computer violates conversational and social conventions will be minimized. The use of whimsical conversation and nonsense, for instance, helps dismiss attempts to entertain conversations on topics that would not be handled sufficiently well by the program. The move from conversations on specific topics to unrestricted conversations after the 1995 edition of the Loebner Prize competition did not reduce the importance of this tactic. On the contrary, nonsense and erratic responses became even more vital for shielding against queries that might have jeopardized the recognition effect.[65] Contestants for the Loebner Prize need to struggle not to create an illusion but, more precisely, to keep it, and this is why the most successful chatbots have applied a "defensive" strategy—finding tricks and shortcuts to restrict the inquiry of the judges, instead of cooperating with them.[66] This approach has also informed attempts to produce effects of personality and characterization in the programming of chatbots for the prize.

GIVING LIFE AT THE LOEBNER PRIZE COMPETITION: CHARACTERIZATION, PERSONALITY, AND GENDER

The winner of the 1994 Loebner Prize was a system called TIPS. Its creator, programmer Thomas Whalen, was driven by a more ambitious vision than those of previous winners, for example, the creators of PC Therapist. His goal was not to confuse judges with nonsense or whimsical conversation but to instill his chatbot with the simple model of a human being, including a personality, a personal history, and a worldview. In the hope of restricting the scope of the conversation, he decided on a male character with a limited worldview who did not read books and newspapers, worked nights, and was therefore unable to watch prime-time television. Whalen also created a story to be revealed gradually to the judges: TIPS's character worked as a cleaner at the University of Eastern Ontario, and on the day of the prize "he" had been concerned about being accused by "his" boss of having stolen something. "I never stole nothing in my life," TIPS stressed in conversations with the judges, "but they always blame the cleaners whenever anything is missing."[67]

Whalen's approach suggests that the problem of creating socially credible chatbots is not only technological but also dramaturgical.[68] Playwrights and novelists, in fact, create characters by feeding audiences and readers aspects of their personal history and their personality. Chatbot developers and AI scientists have often employed literary and theatrical frameworks to reflect on users' interactions with AI agents.[69] For instance the concept of suspension of disbelief—that is, the idea that in consuming a product of fiction we pass through a momentary removal of our normal level of skepticism, so that we respond emotionally to events and stories that we know are fictional—is sometimes mentioned in computer science literature in regard to the effects on users of interactive computer programs, including chatbots.[70] As in conversations with TIPS, in literary fiction characters are often built through a few sentences and actions.[71] In this sense, the play of deception at the Loebner Prize competition can also be seen in terms of characterization and storytelling.

One further complication, however, is the dialogical dimension in which characterization emerges at the Loebner Prize competition. As semiotic theory has shown, readers of fiction actively participate in the construction of meaning.[72] Yet characterization through a chatbot largely differs from fiction, involving its own specific dynamics. The output, in fact, results from a combination of the chatbot's programming and the particular exchanges led by its interlocutor. In the setting of the prize, what a judge contributes to the conversation informs which lines and responses are provided by the chatbot.[73] The scripting of the chatbot's character therefore needs to take into account how different judges could lead the conversation—something that proved very difficult to forecast.[74]

The implications of this uncertainty can be observed in the case of TIPS. Whalen's strategy was largely effective, but it worked only as long as the judges remained within the narrow margins of the chatbot's storyline. In fact, when in 1995 the Loebner Prize regulations regarding topics changed and conversations became unrestricted, the defending champion entered competition with an evolution of TIPS, Joe the Janitor, but the new version of its program only ranked second. Whalen had some quite interesting reflections about the reasons for his defeat:

> First, I had hypothesized that the number of topics that would arise in an open conversation would be limited. . . . My error was that the judges, under Loebner's rules, did not treat the competitors as though they were strangers. Rather, they specifically tested the program with unusual questions like, "What did you have for dinner last night?" or "What was Lincoln's first name?" These are questions that no one would ever ask a stranger in the first fifteen minutes of a

conversation. . . . Second, I hypothesized that, once the judge knew that he was talking to a computer, he would let the computer suggest a topic. . . . Thus, my program tried to interest the judges in Joe's employment problems as soon as possible. . . . I was surprised to see how persistent some judges were in refusing to ever discuss Joe's job.[75]

What emerges in Whalen's words is the difficulty of using chatbots as storytelling devices. Judges did not accommodate Joe the Janitor's storyline; they resisted having a scripted narrative told to them. This is due at least partially to the particular platform of the Loebner Prize competition, an arena of deception and suspicion where judges struggle to expose their conversation partners and actively question their artificial nature. Yet something of this dynamic also informs attempts to instill personality in chatbots in other contexts.

Gina Neff and Peter Nagy have argued that bot agency in social media is built through a symbiotic relationship between the computer program and the user. This applies particularly well in the case of Tay, the machine learning bot released by Microsoft on Twitter in 2016 to "learn" how to "talk" with users through sustained interactions with Twitter users. (The bot was soon discontinued, as it started to imitate racist and abusive messages that had been directed to it by users.) Yet Neff and Nagy's claim that interactions with and reactions to bots are "part of the social and technological apparatus that supports agency within these systems" applies to all categories of AI conversation agents.[76] Although the degree to which and the ways this happens are specific to each platform, personality in bots is an effect that relies on the active contribution of human users. And this is contribution on many different levels: on the one side the user's inputs inform the direction taken by the chatbot's script; on the other the user is also expected to project meaning and personality onto the artificial agent.[77]

This is why attempts to make chatbots communicate with each other often look so weird. Take, for instance, the famed conversation between ELIZA and PARRY that was staged by Vint Cerf in 1973 for a demonstration at a computer conference:

[ELIZA] Would you like to discuss your nerves?
[PARRY] *I guess so.
What does that suggest to you?
*What are you getting at?
What answer would please you most?
*It's not a matter or like or dislike.
What else would you like to discuss?

*I am getting tired of answering the same questions over and over.
What resemblance do you see?
*Why do you need to know that?
What do you think?
*I don't understand your motives.
Why don't you understand my motives?[78]

This conversation shows the limitations of these early chatbots, as well as the incompatibility of ELIZA and PARRY's rules: the routines of ELIZA were planned to deflect while PARRY was programmed to provoke.[79] But this conversation also makes it evident that the chatbots were programmed to enter into conversations with humans rather than with other machines. It is useful, in this sense, to recall the discussion in chapter 2 of McLuhan's argument that media are "extensions" of humans. Modern media are created in the image of humankind: they are envisioned, developed, and fabricated so that they can adapt to their users. Chatbots, and AI agents planned for communication with humans more generally, are no exception. They are designed with a model of the human in mind, which also takes into account the motivation of this imagined user in the expected context of interaction. This model might be flawed, as shown by Whalen's comments about his failure to foresee the behavior of judges at the Loebner Prize competition. Moreover, this model is often restricted to a narrow vision of the human that neglects alterity of race, gender, and class. But it certainly directs the programmers' choices in characterizing their chatbots. In this sense, the fact that chatbots converse poorly with each other is not because to have chatbots do so is impossible or even complicated at a technical level but simply because they are not meant to. The very idea of creating an effect of personality in chatbots, in fact, entails the recognition that it will be achieved through the contribution—in terms of projection and attribution of sense—of human users.

Characterization is, in this regard, first and foremost in the eyes of the beholder. Humans, in fact, are accustomed to attributing agency and personality. They do so through experiences in their everyday life, as well as through their consumption of fiction when they encounter fictional characters in a story. This is manifest in the way chatbots at the Loebner Prize competition, just like other media, such as television or the press, manipulate representations of race, gender, and class to build characters that judges will deem credible.[80] Mark Marino argues that successful performances of chatbots are "the result of conformity between the expectations of appropriate gendered behavior of the interactor and the interactions with the chatbot."[81] These expectations did not go unanticipated by Loebner Prize competitors, as is evident in the varied approaches to drawing on stereotyped representations of gender and sexism. In the second year of the competition, for instance, programmer Joseph Weintraub chose the topic "men vs women" for the restricted Turing test.[82] In 1993, Ken Colby and his son Peter chose the topic "bad marriage"; their chatbot presented itself as a woman and complained, with blatantly sexualized lines, that "my husband is impotent and im [*sic*] a nymphomaniac."[83] A notable example of a chatbot that adopted sexist stereotypes is Julia, which was initially developed to inhabit the online community tinyMUDs and later competed for the Loebner Prize. This chatbot appealed to menstruation and premenstrual syndrome in the attempt to justify its inconsistent responses through gender stereotypes.[84]

Originally, Turing introduced his test as a variation of the Victorian game in which players had to find out whether their interlocutor was male or female.[85] This suggests, as some have noted, that the Turing test is positioned at the intersection between social conventions and performances.[86] As a person who experienced intolerance and was later convicted for his sexual behaviors, Turing was instinctively sensitive to gender issues, and this might have informed his proposal, consciously or unconsciously.[87] In the Loebner Prize competition, the adoption of gender representations as a strategy further corroborates the hypothesis that chatbots are extensions of humans, as they adapt to and profit from people's biases, prejudices, and beliefs. Because judges are willing to read others according to conventional cues, the performance of social conventions and socially held beliefs is meant to strengthen the effect of realism that programmers are seeking. This applies to gender as well as to other forms of representations, such as race (recall the case of the Russian bot who tricked Epstein) and class (as in the story of the cleaner invented by Whalen to win the prize).[88]

In the arena of deception staged at the Loebner Prize competition, gender stereotypes are employed as a tactic to deceive judges. However, chatbots are also programmed to perform gender in other platforms and systems. Playing with gender-swapping has been a constant dimension of online interactions, from multi-user dungeons in the 1980s to chatrooms and to present-day social media. Similar strategies have also been replicated in the design of social interfaces, some of which enact familiar scripts about women's work, for instance in the case of customer services.[89] This was also the case of Ms. Dewey, a social interface developed by Microsoft to overlay its Windows Live Search platform between 2006 and 2009. Represented as a nonwhite woman of ambiguous ethnic identity, Ms. Dewey was programmed to react to specific search terms with video clips that contained titillating and gendered content. As Miriam Sweeney convincingly shows in her doctoral dissertation on the topic, gender and race served as infrastructural elements that facilitated ideological framing of web searches in this Microsoft platform.[90] The Ms. Dewey interface thus reinforced through characterization the underlying bias of search engine algorithms, which have often been found to reproduce racism and sexism and orient users toward biased search results. Ms. Dewey, in other words, reproduced at a representational level, with the video clips, what the information retrieval systems giving access to the web executed at an operational level.[91] This approach reverberates in the characterization of contemporary AI voice assistants, such as Alexa, which is offered by default to customers as a female character.[92]

THE LOEBNER PRIZE AND ARTIFICIAL SOCIALITY

To date, no computer program that has competed for the Loebner Prize has passed the Turing test. Even if one ever does so, though, it will not mean that a machine is "thinking" but, more modestly, as Turing himself underlined in his field-defining article, only that a computer has passed this test. Given the hype that has surrounded the prize, as well as other human-versus-computer contests in the past, one can anticipate that such a feat would nonetheless enhance perceptions of the achievements of AI, even if it might not have an equivalent impact at the level of technical progress. Indeed, the popularity of such events attests to a fear, but also a deep fascination, in contemporary cultures with the idea of intelligent machines. This fascination is evident in fictional representations of AI: the dream of artificial consciousness and the nightmare of the robot rebellion have informed much science fiction film and literature since the emergence of computers. But this fascination is also evident in reports and discussions on public performances where AI is celebrated as a form of spectacle, like the Loebner Prize competition. People seem to have a somehow ambivalent, yet deep, desire to see the myth of the thinking machine come true, which shapes many journalistic reports and public debates about the Turing test.[93]

As I have shown, the Loebner Prize competition says more about how people are deceived than about the current state of computing and AI. The Loebner Prize competition has become an arena of deception that places humans' liability for being deceived at center stage. Yet this does not make this contest any less interesting and useful for understanding the dynamics of communicative AI systems. It is true that in "real life," which is to say in the most common forms of human-computer interaction, users do not expect to be constantly deceived, nor do they constantly attempt to assess whether they are interacting with a human or not. Yet it is becoming harder and harder to distinguish between human and machine agents on the web, as shown by the diffusion of social media bots and the ubiquity of CAPTCHAs—which are reversed Turing tests to determine whether or not the user making a request to a web page is a human.[94] If it is true that the possibility of deception is heightened by the very limits the Loebner Prize competition places on communicative experiences, such as the textual interface and the prize's regulations, it is also true that similar "limits" characterize human-machine communication on other platforms. As Turkle points out, "even before we make the robots, we remake ourselves as people ready to be their companions." Social media and other platforms provide highly structured frameworks that make deception relatively easier than in face-to-face interactions or even in other mediated conversations, as in a phone call.[95]

The outright deceptions staged at the Loebner Prize competition, moreover, help one to reflect on different forms of relationships and interactions with AI that are shaped by banal forms of deception. As will be discussed in the next chapter, even voice assistants such as Alexa and Siri have adopted some of the strategies developed by Loebner Prize contestants—with the difference that the "tricks" are aimed not at making users believe that the vocal assistant is human but, more subtly, at sustaining the peculiar experience of sociality that facilitates engagements with these tools. And bots in social media, which have recently been used also for problematic applications such as political communication, have appropriated many of the strategies illuminated by the Loebner Prize competition.[96] Among the tricks employed by competitors in the prize that have been integrated into the design of AI tools of everyday use are turn taking, the slowing of responses in order to adapt to the rhythms of human conversation, and the use of irony and banter. Injecting a personality and creating a convincing role for the chatbot in a conversation has been instrumental to success at the prize but is also a working strategy to create communicative AI systems that can be integrated into complex social environments such as families, online communities, and professional spaces.[97]

The Loebner Prize competition, in this sense, is an exercise in simulating sociality as much as in simulating intelligence. Playing the Turing test successfully goes beyond the capacity to grasp the meaning of a text and produce an appropriate response. It also requires the capacity to adapt to social conventions of conversation and exchanges. For example, because emotions are an integral part of communications between humans, a computer program benefits from being capable of recognizing emotions and devising appropriately affective responses.[98] While other human-versus-machine competitions, such as chess, emphasized the machine's capacity of making strategic decisions, the Loebner Prize competition makes sociality a defining feature for AI. Its key message is not that AI will or will not ever be achieved but rather that there is nothing like "artificial intelligence": there are only socially minded interactions that lead to attributing intelligence to machines.

CHAPTER 6

To Believe in Siri

A Critical Analysis of Voice Assistants

"Talk to Siri as you would to a person," suggested Apple to the users of its voice assistant Siri in 2011, after it was bundled into the iPhone operating system.[1] This message was meant to inspire a sense of familiarity with the assistant. Apple suggested that everything was already in place to accommodate the new technology in everyday experience: users just needed to extend their conversational habits to the invisible interlocutor embedded in the phone.

Given the swift success of Siri and other voice assistants in the following years, Apple's incitation may have worked. Similar tools were soon developed by other leading digital corporations: Amazon introduced Alexa in 2014, Google followed with its Assistant in 2016, while Microsoft's Cortana, now being discontinued, was launched even earlier, in 2013. In just a few years, the technology left the confined spaces of smartphones to dwell in all sorts of digital devices, from watches to tablets and speakers, inhabiting both domestic and professional environments. Just as graphic interfaces draw on visual information to facilitate interaction, voice assistants are based on software that recognizes and produces voice inputs. Users' commands and questions are then elaborated through language-processing algorithms that provide replies to the users' queries or execute tasks such as sending emails, searching on the web, and turning on lamps. Each assistant is represented as an individual character or persona (e.g., "Siri" or "Alexa") that despite being nonhuman can be imagined and interacted with as such. As confirmed by market research and independent reports, they have been

Deceitful Media. Simone Natale, Oxford University Press (2021). © Oxford University Press.
DOI: 10.1093/oso/9780190080365.003.0007

adopted by hundreds of millions of users around the world, making voice a key medium of interaction with networked computer technologies.[2]

The incitation of companies such as Apple to talk to voice assistants "as to a person," however, deserves to be questioned. Have voice assistants developed into something we talk to "as we would to a person," as promised by Siri's marketing lines? And if so, what does this even mean? Focusing on the cases of Alexa, Siri, and Google Assistant, this chapter argues that voice assistants activate an ambivalent relationship with users, giving them the illusion of control in their interactions with the assistants while at the same time withdrawing them from actual control over the computing systems that lie behind these interfaces. I show how this is made possible at the interface level by mechanisms of banal deception that expect users to contribute to the construction of the assistant as a persona, and how this construction ultimately conceals the networked computing systems administered by the powerful corporations who developed these tools.

A critical analysis of voice assistants means unveiling the different strategies and mechanisms by which users are encouraged to accommodate existing social habits and behaviors so that they can "talk" to the AI assistant. Such strategies are not by any means straightforward, and do not correspond to tricking the users into believing that the AI thinks or feels "like a person." Artificial intelligence assistants rely on humans' tendency to project identity and humanity onto artifacts but at the same time do not imply any decision from users regarding their ontology. In other words, they do not require users to decide whether they are talking to a machine or to a person. They require them just to talk.

Although users ultimately benefit from the functionality of AI assistants and an enhanced capacity to accommodate the new technology in their everyday lives, one is left questioning whether it is safe to trust companies such as Apple, Amazon, and Google to micromanage more parts of our lives. The only way to find an answer to this question is by looking through the complex stratification of technologies and practices that shape our relationship with these tools.

ONE AND THREE

In the Christian theological tradition, God is "one and three": Father, Son, Holy Spirit. This doctrine, called the Trinity, has stimulated lively theological discussions across many centuries. It is in fact one of the elements of the Christian faith that appears most confusing to believers: the idea that God's three "persons" are distinct and one at the same time contrasts with widely held assumptions about individuality, by which being one and being three are mutually exclusive.[3]

A similar difficulty also involves software. Many systems that are presented as individual entities are in fact a combination of separate programs applied to diverse tasks. The commercial graphic editor software package known as Photoshop, for instance, hides behind its popular trademark a complex stratification of discrete systems developed by different developers and teams across several decades.[4] When looking at software, the fact that what is one is also at the same time many should be taken not as the exception but as the norm. This certainly does not make software closer to God, but it does make it a bit more difficult to understand.

Contemporary voice assistants such as Alexa, Siri, and Google Assistant are also one and many at the same time. On the one side they offer themselves to users as individual systems with distinctive names and humanlike characteristics. On the other side, each assistant is actually the combination of many interconnected but distinct software systems that perform particular tasks. Alexa, for instance, is a complex assemblage of infrastructures, hardware artifacts, and software systems, not to mention the dynamics of labor and exploitation that remain hidden from Amazon's customers.[5] As BBC developer Henry Cooke put it, "there is not such a thing as Alexa" but only a multiplicity of discrete algorithmic processes. Yet Alexa is perceived as one thing by its users.[6]

Banal deception operates by concealing the underlying functions of digital machines through a representation constructed at the level of the interface. A critical analysis of banal deception, therefore, requires examination of the relationship between the two levels: the superficial level of the representation and the underlying mechanisms that are hidden under the surface, even while they contribute to the construction of the overlaid representation. In communicative AI, the representation layer also coincides with the stimulation of social engagement with the user. Voice assistants draw on distinctive elements such as a recognizable voice, a name, and elements to suggest a distinctive persona such as "Alexa" or "Siri." From the user's point of view, a persona is above all an imagined construction, conveying the feeling of a continuing relationship with a figure whose appearance can be counted on as a regular and dependable event and is integrated into the routines of daily life.[7]

In the hidden layer are multiple software processes that operate in conjunction but are structurally and formally distinct. Although the entire "anatomy" of voice assistants is much more complex, three software systems are crucial to the functioning of voice assistants' banal deception, which roughly map to the areas of speech processing, natural language processing, and information retrieval (fig. 6.1). The first system, speech

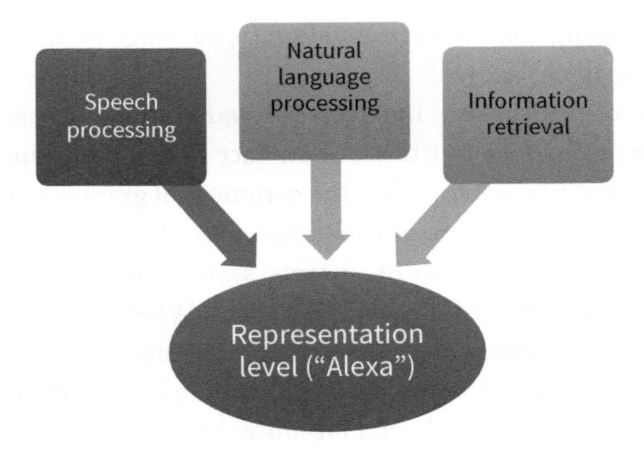

Figure 6.1 AI voice assistants as "one and three."

processing, consists of algorithms that on the one hand "listen to" and transcribe the users' speech and on the other produce the synthetic voice through which the assistants communicate with users.[8] The second system, natural language processing, consists of the conversational programs that analyze the transcribed inputs and, as for a chatbot program, elaborate responses in natural language.[9] The third system, information retrieval, consists of the algorithms that retrieve relevant information to respond to the users' queries and activate the relevant tasks. Compared to speech processing and natural language processing, the relevance of information retrieval algorithms is perhaps less evident at first glance. However, they enable voice assistants to access Internet-based resources and to be configured as users' proxies for navigating the web.[10] As the following sections will show, the differences between these three software systems are not restricted to their functions, since each of them is grounded in distinct approaches to computing and AIs, and carries with it different implications at both the technical and social levels.

SPEECH PROCESSING, OR THE SOFT POWER OF VOICE

Since the invention of media such as the phonograph and the telephone, scientists and engineers have developed a range of analog and digital systems to record and reproduce sound. Like all modern media, sound media, as discussed earlier, were produced in the shape of the human user. For example, studies of human hearing were incorporated into the design of technologies such as the phonograph, which recorded and reproduced sound that matched sound frequencies perceived by humans.[11] A similar work of adaptation also involved the human voice, which was immediately envisaged as the key field of application for sound reproduction and recording.[12] For instance, in 1878 the inventor of the phonograph, Thomas Alva Edison, imagined not music but voice recordings for note-taking or family records as the most promising applications for its creation.[13] Specific efforts were therefore made to improve the mediation of the voice.

Such endeavors profited from the fact that people are built or "wired" for speech.[14] A human voice is more easily heard by humans than other noises, and familiar voices are recognized with a precision barely matched by the vision of a known face. This quality comes to fruition in mediated communications. In cinema, for instance, the tendency of audiences to recognize a voice and immediately locate its source accomplishes several functions, adding cohesion to the narrative and instilling the sense of "life"—that is, presence and agency—to characters.[15] Similarly, in voice conversations over the phone or other media, the capacity to recognize and identify a disembodied voice is essential to the medium's use. This skill enables phone users to recognize a familiar person and to gain hints about the demographics and mood of a voice's owner. Thus the technical mediation of voice draws on the characteristics of human perception to generate meaningful results that are fundamental to the experience of the user.

Following from this technical and intellectual lineage, the dream of using spoken language as an interface to interact with computers is as old as computing itself.[16] Until very recently, however, voice-based interfaces struggled to provide reliable services to users. Encountering automatic voice recognition technologies was often a frustrating experience: early "press or say one" phone services didn't handle variations in accent or tone well, and users were often annoyed or amused by these systems' struggles to comprehend even simple inputs.[17] Compared with the performances of Alexa or Siri, especially but not exclusively in the English language, one wonders how the technology improved so markedly in such a short lapse of time. The secret of this swift progress lies in one of the most significant technical changes that AI has experienced throughout its history: the rise of deep learning.

Deep learning refers to a class of machine-learning algorithms that rely on complex statistical calculations performed by neural networks autonomously and without supervision. Inspired by the functioning of biological neurons, neural networks were proposed very early in the history of AI but initially seemed ineffective.[18] In the 1980s and 1990s, new studies showed at a theoretical level that neural networks could be extremely powerful.[19] Yet only in the last decade has the technology realized its full potential, due to two main factors: advances in hardware that have made computers able to process the complex calculations required by neural networks and, even more important, the availability of huge amounts of data, often produced by human users on the Internet, to "train" deep learning algorithms for the performance of specific tasks.[20]

More broadly, deep learning has emerged in conjunction with a recalibration of human-computer interactive systems. For a growing range of AI applications, "intelligent" skills are not programmed symbolically into the machine. Instead, they emerge through statistical elaboration of human-generated data that are harvested in computing networks through new forms of labor and automated power relations.[21] Together with other applications, such as image analysis and automatic translation, speech processing is one of the areas of AI that has most benefited from the rise of deep learning. In the span of just a few years, the availability of masses of data that could be used to "train" the algorithms catalyzed a jump ahead in the automatic processing of the human voice by computers. Speech processing was in this sense the veritable killer application of AI assistants.[22]

As the technical processing of the human voice became more sophisticated, companies that developed voice assistants took great care to adapt speech processing to their target users—exactly as analog sound media profited from studies about the physiology and psychology of their audiences to improve the recording and reproduction of the voice.[23] Significant thoughts were given to calibrating the way the assistant's synthetic voice would sound to the ears of human users and to anticipate potential reactions to specific modulations.

Apple's Siri initially employed three different voiceover artists to represent the United States, Australia, and the United Kingdom.[24] This decision was a response to the need to adapt to different English accents, as well as to accommodate what Apple developers thought were cultural differences regarding perceptions of male versus female voices—hence the decision to employ a male voice in the United Kingdom and a female voice for the United States and Australia.[25] Later, Apple included further accents for the English language, such as Irish and South African, and—also in response to controversies about gender bias—allowed customization of gender. Google Assistant launched with a female default voice but introduced a number of male voices and opted, also in reaction to controversies about sexism, for a random selection of one of the available voices as default.[26] Alexa is clearly characterized as female, although customization options have been recently integrated by Amazon, which also launched a remarkable new add-on by which the voice of American actor Samuel Lee Jackson is made available to Alexa users for the affordable price of $0.99.[27]

The fact that voice assistants have often featured female voices as default has been at the center of much criticism. There is evidence that people react to hints embedded in AI assistants' voices by applying categories and biases routinely attributed to people. This is particularly worrying if one considers the assistants' characterization as docile servants, which reproduces stereotypical representations of gendered labor.[28] As Thao Phan has argued, the representation of Alexa directs users toward an idealized vision of domestic service, departing from the historical reality of this form of labor, as the voice is suggestive of a native-speaking, educated, white woman.[29] Similarly, Miriam Sweeney observes that most AI assistants' voices suggest a form of "'default whiteness' that is assumed of technologies (and users) unless otherwise indicated."[30]

Although public controversies have stimulated companies to increase the diversity of synthetic voices, hints of identity markers such as race, class, and gender continue to be exploited to trigger a play of imagination that relies on existing knowledge and prejudice.[31] Studies in human-computer communication show that the work of projection is performed automatically by users: a specific gender, and even a race and class background, is immediately attributed to a voice.[32] The notion of stereotyping, in this sense, helps us understand how AI assistants' disembodied voices activate mechanisms of projection that ultimately regulate their use. As Walter Lippmann showed in his classic study of public opinion, people could not handle their encounters with reality without taking for granted some basic representations of the world. In this regard, stereotypes have ambivalent outcomes: on the one side they limit the depth and detail of people's insight into the world; on the other they help people recognize patterns and apply interpretative categories built over time. While negative stereotypes need to be exposed and dispelled, Lippmann's work also shows that the use of stereotypes is essential to the functioning of mass media, since knowledge emerges both through discovery and through the application of preconstituted categories.[33]

This explains why the main competitors in the AI assistant market select voices that convey specific information about a "persona"—that is, a representation of individuality that creates the feeling of a continuing relationship with the assistant itself. In other computerized services employing voice processing technologies, such as customer services, voices are sometimes made to sound neutral and evidently artificial. Such a choice, however, has been deemed untenable by companies who aspire to sell AI assistants that will accompany users throughout their everyday lives. To function properly, these tools need to activate the mechanisms of representation by which users imagine a source for the voice—and, subsequently, a stable character with whom to interact, even if within relatively strict boundaries.[34] As Lippmann has taught us, such mechanisms rely on previous stereotypes—which makes the choice of a gendered voice strategic, if extremely problematic, for companies such as Apple and Amazon.[35]

Voice assistants encourage users, by relying on them to apply their own stereotyping, to contribute actively to the construction of sense around the disembodied voice. This does not serve much to give "life" to the assistants—despite all the anthropomorphic cues, users retain the ability to differentiate clearly between AI assistants and real persons.[36] More subtly, stereotyping helps a user assign a coherent identity to an assistant over time. This is achieved in the first instance by making the assistant's voice recognizable: a voice assistant can only be assigned a coherent personality by the user due to the fact that its voice always sounds the same.[37] The attribution of gender, race, and class through stereotyping prompts further signals to nurture the play of imagination involved in the user's construction of a persona.[38]

Thus the synthetic, humanlike voices of AI assistants, as anthropomorphic cues, are not meant to produce the illusion of talking to a human but to create the psychological and social conditions for projecting an identity and, to some extent, a personality onto the virtual assistant. This banal form of deception does not imply any strict definition on the user's part: one can grasp perfectly that Alexa is "just" a piece of software and at the same time carry out socially meaningful exchanges with it. As it is ultimately left for the users to attribute social meaning, voice assistants leave ample space for individual interpretation.

This helps explain why research has shown that people construct their relationships with voice assistants in very diverse ways. For instance, in Andrea Guzman's qualitative research with users of mobile conversational agents, participants gave a range of interpretations of aspects such as a voice's source, some of them identifying it as a voice "in" the mobile phone and others perceiving it as the voice "of" the phone. The fact that "the locus and nature of the digital interlocutor is not uniform in people's minds" is a result of the high degree of participation that is required from the user.[39] Likewise, recent research shows that different users retain diverse types of benefits from their interactions with the assistants.[40] To use McLuhan's term, voice assistants are a "cool" medium, a notion McLuhan applies to media such as television and the telephone that are low-definition and require participation on the part of the audience.[41] The low definition of Alexa and other AI assistants leaves listeners to do the bulk of the work. As a result, though, Alexa is different things to different kinds of users. It is, after all, a necessity of any medium of mass diffusion to be able to adapt its message to diverse populations.

Still, the design of voice interfaces exercises an undeniable influence over users. Experimental evidence shows that synthetic voices can be manipulated to encourage the projection of demographic cues, including gender, age, and even race, as well as personality traits, such as an extroverted or an introverted, a docile or aggressive character.[42] Yet this is a ultimately a "soft," indirect power, whereby the attribution of personality is delegated to the play of imagination of the users. The low definition of voice assistants contrasts with humanoid robots, whose material embodiment reduces the space of user contribution, as well as with graphic interfaces, which leave less space for the imagination.[43] The immaterial character of the disembodied voice should not to be seen, however, as a limitation: it is precisely this disembodiment that forces users to fill in the gaps and make voice assistants their own, integrating them more deeply into their everyday lives and identities. As a marketing line for Google Assistant recites, "it's your own personal Google."[44] The algorithms are the same for everybody, but you are expected to put a little bit of yourself into the machine.

OF HAIKUS AND COMMANDS: NATURAL LANGUAGE PROCESSING AND THE DRAMATURGY OF VOICE ASSISTANTS

When I pick up my phone and ask Siri if it's intelligent, it has an answer (fig. 6.2).

To me, this is not just a turn of phrase. It's an inside joke that points to the long tradition of reflections about machines, intelligence, and awareness—all the way back to Turing's 1950 article in which he argued that the question whether machines can "think" was irrelevant.[45] Yet what at first glance looks like a clever reply is actually one of the least "smart" things Siri can do. This reply, in fact, is not the result of a sophisticated simulation of symbolic thought, nor has it emerged from statistical calculations of neural networks. More simply, it was manually added by some designer at Apple who decided that a question about Siri's intelligence should ignite such a reply. This is something a programmer would hesitate to describe as coding, dismissing it as little more than script writing—in the Loebner Prize's jargon, just another "programming trick."

The recent swift success of voice assistants has led many researchers and commentators to argue that these tools have clearly outrun the conversational skills of earlier conversational programs.[46] Yet in contrast with this widely held assumption, the likes of Alexa, Siri, and Google Assistant do

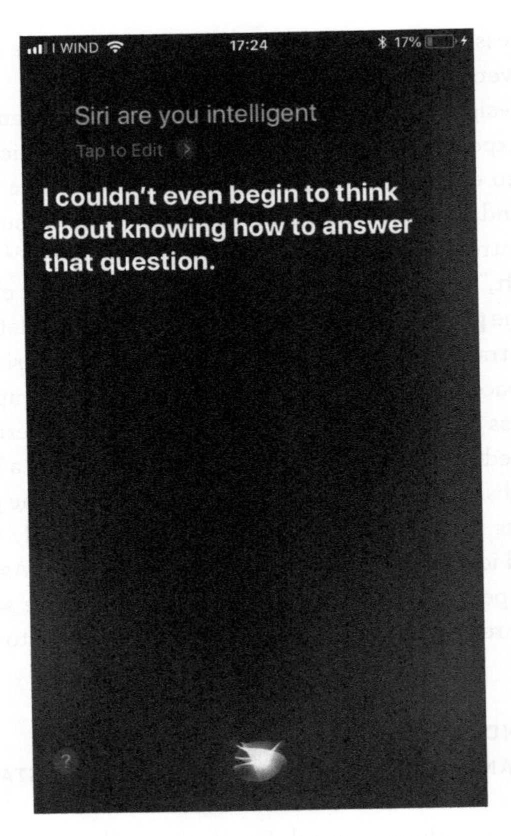

Figure 6.2 Author's conversation with Siri, 15 December 2019.

not depart dramatically from earlier systems, including chatbots, at least in terms of their conversational skills. This is due to the uneven degree of progress AI has experienced in recent years. The rise of deep learning stimulated new expectations for the field and captured the public imagination with the vision that AI would equal or outrun humans in every kind of task in the near future. The myth of the thinking machine has been reignited, and the AI enterprise is experiencing a new wave of enthusiastic responses in the public sphere as well as in scientific circles.[47] The picture, however, is much more complex. For all the potential of neural networks, not all areas of AI have benefited from the deep learning revolution. Due to the difficulty of retrieving and organizing data about conversations and thus the difficulty of training algorithms to this task, conversational systems have until now only been slightly touched by deep learning. As technology journalist James Vincent put it, "machine learning is fantastic at learning vague rules in restricted tasks (like spotting the difference between cats and dogs or identifying skin cancer), but it can't easily turn a stack of data into the complex, intersecting, and occasionally irrational guidelines that make up modern English speech."[48]

Thus, while voice assistants represent a step ahead in communicative AI for areas such as voice processing, their handling of conversations still relies on a combination of technical as well as dramaturgical solutions. Their apparent proficiency is the fruit of some of the same strategies developed by chatbot developers across the last few decades, combined with an unprecedented amount of data about users' queries that helps developers anticipate their questions and design appropriate responses. The dramaturgical proficiency instilled in voice assistants at least partially compensates for the technical limitations of existing conversational programs.[49]

In efforts to ensure that AI assistants reply with apparent sagacity and appear able to handle banter, Apple, Amazon, and to a smaller degree Google assigned the task of scripting responses to dedicated creative teams.[50] Similar to Loebner Prize chatbots, which have been programmed to deflect questions and restrict the scope of the conversations, scripted responses allow voice assistants to conceal the limits of their conversational skills and maintain the illusion of humanity evoked by the talking voice. Every time it is asked for a haiku, for instance, Siri comes out with a different piece of this poetry genre, expressing reluctance ("You rarely ask me / what I want to do today / Hint: it's not haiku"), asking the user for a recharge ("All day and night, / I have listened as you spoke. / Charge my battery"), or unenthusiastically evaluating the genre ("Haiku can be fun / but sometimes they don't make sense. / Hippopotamus").[51] Although the activation of scripted responses is exceedingly simple on a technical level, their ironic tone can be striking to users, as shown by the many web pages and social media posts reporting some of the "funniest" and "hilariously honest" replies.[52] Irony has been instrumental in chatbots gaining credibility at the Loebner Prize, and voice assistants likewise benefit from the fact that irony is perceived as evidence of sociability and sharpness of mind.

In contrast to Loebner Prize chatbots, the objective of voice assistants is not to deceive users into believing they are human. Yet the use of dramaturgical "tricks" allows voice assistants to achieve subtler but still significant effects. Scripted responses help create an appearance of personalization, as users are surprised when Siri or Alexa reply to a question with an inventive line. The "trick" in this case is that AI assistants are also surveillance systems that constantly harvest data about users' queries, which are transmitted and analyzed by the respective companies. As a consequence, AI assistant developers are able to anticipate some of the most common queries and have writers come out with appropriate answers.[53] The consequentiality of this trick remains obscure to many users, creating the impression that the voice assistant is anticipating the user's thoughts—which meets expectations of what a "personal" assistant is for.[54] Users are thereby swayed into believing AI assistants to be more capable of autonomous behavior than they actually are. As noted by Margaret Boden, they appear "to be sensitive not only to topical relevance, but to personal relevance as well," striking users as "superficially impressive."[55]

Also unlike Loebner Prize chatbots, voice assistants have simulated sociality as just one of the functions, not their raison d'être. Social engagement is never imposed on the user but occurs only if users invite this behavior through specific queries. When told "Goodnight," for instance, Alexa will reply with scripts including "Goodnight," "Sweet dreams," and "Hope you had a great day." Such answers, however, are not activated if users just request an alarm for the next morning. Depending on the user's input, AI assistants enact different modalities of interaction. Alexa, Google Assistant, and Siri can be a funny party diversion one evening, exchange conviviality at night, and the next day return to being discreet voice controllers that just turn lights on and off.[56]

This is what makes AI assistants different from artificial companions, which are software and hardware systems purposely created for social companionship. Examples of artificial companions include robots such as Jibo, which combines smart home functionality with the appearance of empathy, as well as commercial chatbots like Replika (fig. 6.3), an AI mobile app that promises users comfort "if you're feeling down, or anxious, or just need someone to talk to."[57] While Alexa, Siri, and Google Assistant are only meant to play along if the user wants them to, artificial companions are purportedly programmed to seek communication and emotional engagement. If ignored for a couple of days, for instance, companionship chatbot Replika comes up with a friendly message, such as "Is everything OK?" or the rather overzealous "I'm so grateful for you and the days we have ahead of us."[58] Alexa and Siri on the other hand require incitement to engage in pleasantries or banter—coherently with one of the pillars of their human-computer interaction design, by which assistants speak up only if users pronounce the wake word. This is also why the design of smart speakers that provide voice assistant services in domestic environments, such as Amazon Echo, Google Home, and Apple HomePod, is so minimal. The assistants are meant to remain seamless, always in the background, and quite politely intervene only when asked to do so.[59]

When not stimulated to act as companions, voice assistants treat conversational inputs as prompts to execute specific tasks. Conversations with AI assistants therefore profit from the fact that language, as shown by speech

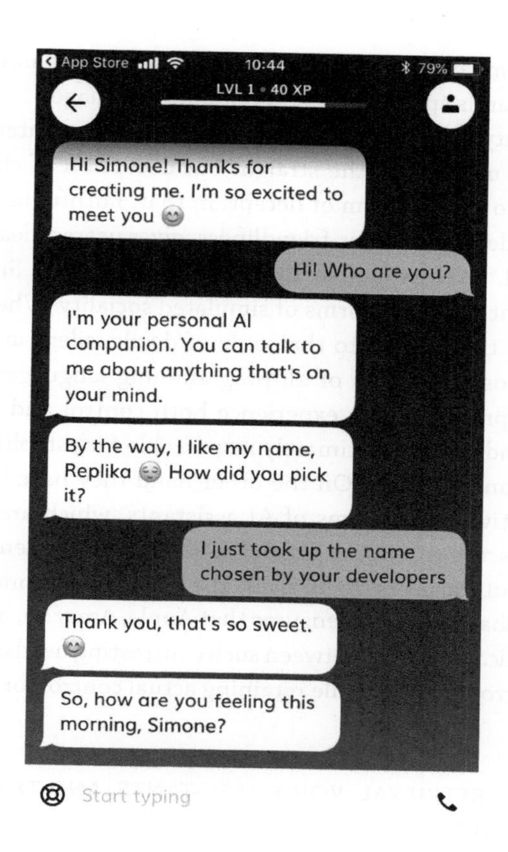

Figure 6.3 Author's conversation with Replika, 30 December 2019. Replika boasts of a reported 2.5 million sign-ups and a community of around 30,000 users on its Facebook group. Replika allows customization of gender (female, male, and nonbinary) and is programmed to be overly submissive and complimentary to the user, its stated purposes being: "Talk things through together," "Improve mental wellbeing," "Explore your personality" and "Grow together." (Luka Inc., "Replika," 2019, available at https://replika.ai/, retrieved 30 December 2019.)

act theory, is also a form of action.[60] As mentioned in chapter 3 in regard to dialogue trees, some have proposed that natural language is equivalent to a very high-level programming language.[61] It is easy to see how this applies in the case of communicative AI interfaces such as the assistants.[62] In computing, every line of code is translated into lower-level specific commands, all the way down to the physical operations performed at the level of the machine. Similarly, when users ask Siri or Alexa to "call Mom" or "play radio" these inputs are translated into the corresponding instructions in lower-level programming languages to initiate corresponding functions.[63] An expert user of voice assistants will learn which commands most effectively get Alexa, Siri, or Google Assistant to "understand" and operate accordingly—similar to the way a computer scientist memorizes the most frequent commands specific to a programming language.

The variability of voice assistants' approaches to interaction is exemplary of the move from the straight-out deception of chatbots in the Loebner Prize to a banal form of deception that, faithful to the principles of transparent design and user-friendliness, gives users at least the appearance of control.[64] When ignited by appropriate queries, Siri, Alexa, and Google Assistant engage in forms of simulated sociality. When approached with a request, they return to their role of docile, silent aides controlled through the most accessible of all programming languages: natural language. In this process, users experience both control and the lack of it. On the one hand they are ultimately responsible for establishing the tone and scope of conversations. On the other hand they have limited insight into the deceptive mechanisms of AI assistants, which are embedded in code that is obscure to most users and relies on unprecedented degrees of knowledge about the users themselves. The incessant accumulation of data about users' behavior, in fact, ensures that Apple, Amazon, and Google can manage the delicate balance between such contrasting needs: conceding the illusion of control to users while retaining actual control for themselves.

INFORMATION RETRIEVAL, VOICE ASSISTANTS, AND THE SHAPE OF THE INTERNET

Both speech processing and the handling of conversation are examples of the "low definition" of voice assistants. On the one side the processing of voice requires users to contribute to the construction of the assistant's persona through stereotyping. On the other side the conversational routines of AI assistants adapt to the users' queries, offering different modalities of interaction and leaving users with the illusion of control over the experience. In contrast, the third system—information retrieval—has more to do with the tasks voice assistants are able to complete than with users' perceptions of computing systems. Yet there is a close link between the apparently neutral and mundane character of information retrieval operations and the overall representation level of each assistant, which in turn helps one understand how users are withdrawn control over the networked computing systems they are interacting with.

The term *information retrieval* refers to systems that enable the localization of information relevant to specific queries or searchers.[65] Information retrieval regulates the functioning of web search engines such as Google Search and, more generally, the retrieval of information across the web.

Little attention, however, has been given to the fact that information retrieval also plays a key role in voice assistants. To properly function, voice assistants such as Alexa, Siri, and Google Assistant need to be constantly connected to the Internet, through which they retrieve information and access services and resources. Internet access allows these systems to perform functions that include responding to queries, providing information and news, streaming music and other online media, managing communications—including emails and messaging—and controlling smart devices in the home such as lights or heating systems.[66] Although voice assistants are scarcely examined as to the nature of the interfaces they use that give access to Internet-based resources, they are ultimately technologies that provide new pathways to navigating the web through the mediation of huge corporations and their cloud services. As voice assistants enter more and more into public use, therefore, they also inform the way users employ, perceive, and understand the web and other resources.

One of the key features of the web is the huge amount of information that is accessible through it.[67] In order to navigate such an imposing mass of information, users employ interfaces that include browsers, search engines, and social networks. Each of these interfaces helps the user identify and connect to specific web pages, media, and services, thereby restricting the focus to a more manageable range of information that is supposed to be tailored to or for the user.

To some extent, all of these interfaces could be seen as empowering users, as they help them retrieve information they need. Yet these interfaces also have their own biases, which correspond to a loss of control on the part of the user. Search engines, for instance, index not all but only parts of the web, influencing the visibility of different pieces of information based on factors including location, language, and previous searches.[68] Likewise, social networks such as Facebook and Twitter impact on access to information, due to the algorithms that decide on the appearance and ranking of different posts as well as to the "filter bubble" by which users tend to become distanced from information not aligning with their viewpoints.[69] It is for this reason that researchers, since the emergence of the web, have kept interrogating whether and to what extent different tools for web navigation facilitate or hinder access to a plurality of information.[70] The same question urgently needs to be asked for voice assistants. Constructing a persona in the interface, voice assistants mobilize specific representations while they ultimately reduce users' control over their access to the web, jeopardizing their capacity to browse, explore, and retrieve a plurality of information available through the web.

A comparison between the search engine Google and the voice assistant Google Assistant is useful at this point. If users search one item on the search engine, say "romanticism," they are pointed to customized entries from Wikipedia and the Oxford English Dictionary alongside a plethora of other sources. Although studies show that most users rarely go beyond the first page of a search engine's results, the interface still enables users to browse at least a few of the 16,400,000 results retrieved through their search.[71] The same input given to Google Assistant (at least, the version of Google Assistant on my phone) linked only to the Wikipedia page for "Romanticism," the artistic movement. The system disregards other meanings for the same words in the initial search and privileges one single source. If the bias of Google algorithms applies to both the search engine and the virtual assistant, in Google Assistant browsing is completely obliterated and replaced by the triumph of "I'm feeling lucky" searches delivering a single result. Due to the time that would be needed to provide several potential answers by voice, the relative restriction of options is to be considered not just a design choice but a characteristic of voice assistants as a medium.

Emily MacArthur has pointed out that a tool such as Siri "restores a sense of authenticity to the realm of Web search, making it more like a conversation between humans than an interaction with a computer."[72] One wonders, however, whether this "sense of authenticity" is a way for voice assistants to appear to be at the service of the users, to make users forget that their "assistants" are at the service of the companies that developed them. In spite of their imagined personae, "Alexa," "Siri," and "Google Assistant" never exist on their own. They exist only embedded in a hidden system of material and algorithmic structures that guarantee market dominance to companies such as Amazon, Apple, and Google.[73] They are gateways to the cloud-based resources administered by these companies, eroding the distinction between the web and the proprietary cloud services that are controlled by these huge corporations. This erosion is achieved through close interplay between the representation staged by the digital persona embodied by each assistant and its respective company's business model.

It is striking to observe that the specific characterization of each voice assistant is strictly related to the overall business and marketing approaches of each company. Alexa is presented as a docile servant, able to inhabit domestic spaces without crossing the boundaries between "master" and "servant." This contributes to hiding Amazon's problematic structures of labor and the precarious existence of the workers who sustain the functionality of the platform.[74] Thus, Alexa's docile demeanor contributes to making the exploitation of Amazon's workforce invisible to the customer-users who access Amazon Prime services and online commerce through Alexa. In turn, Siri, compared to the other main assistants, is the one that makes the most extensive use of irony. This helps corroborate Apple's corporate image of creativity and uniqueness, which Apple attempts to project onto its customers' self-representation: "stay hungry stay foolish," as recited in a famous Apple marketing line.[75] In contrast with Apple and Amazon, Google chose to give their assistant less evident markers of personal identity, avoiding even the use of a name.[76] What appears to be a refusal to characterize the assistant, however, actually reflects Google's wider marketing strategy, which has always downplayed elements of personality (think of the low profile, compared to Steve Jobs at Apple or Jeff Bezos at Amazon, of Google's founders, Larry Page and Sergei Brin) to present Google as a quasi-immanent oracle aspiring to become indistinguishable from the web.[77] Google Assistant perpetuates this representation by offering itself as an all-knowing entity that promises to have an answer for everything and is "ready to help, wherever you are."[78]

Rather than being separated from the actual operations voice assistants carry out, the construction of the assistant's persona is meant to feed into the business of the corporations. In fact, through the lens of Siri or Alexa, there is no substantial difference between the web and the cloud-based services administered by Apple and Amazon.[79] Although interfaces are sometimes seen as secondary to the communication that ensues through them, they contribute powerfully to shape the experiences of users. It is for this reason that the nature of voice assistants' interfaces needs to be taken seriously. Like the metaphors and representations evoked by other interfaces, the construction of the assistant's persona is not neutral but informs the very outcome of the communication. In providing access to the web, voice assistants reshape and repurpose it as something that responds more closely to what companies such as Amazon, Apple, and Google want it to look like for their customers.

TO DECEIVE AND NOT TO DECEIVE

Voice assistants represent a new step toward the convergence between AI and human-computer interaction.[80] As media studies scholars have shown, all interfaces employ metaphors, narrative tropes, and other forms of representation to orient interactions between users and machines toward specific goals.[81] Graphic interfaces, for instance, employ metaphors such as the desktop and the bin, constructing a virtual environment that hides the complexity of operating systems through the presentation of elements familiar to the user. The metaphors and tropes manifested in the interface inform the imaginary constructions through which people perceive, understand, and imagine how computing technologies work.[82]

In voice assistants, the level of representation coincides with the construction of a "persona" that reinforces the user's feeling of a continuing relationship with the assistant. This adds further complexity to the interface, which ceases to be just a point of intersection between user and computer and takes on the role of both the channel and the producer of communication. The literal meaning of *medium*, "what is in between" in Latin, is subverted by AI assistants that reconfigure mediation as a circular process in which the medium acts simultaneously as the endpoint of the communication process. An interaction with an AI assistant, in this sense, results in creating additional distance between the user and the information retrieved from the web through the indirect management of the interface itself.

This chapter has shown the way this distancing is created through the application of banal deception to voice assistants. Mobilizing a plurality of technical systems and design strategies, voice assistants represent the continuation of a longer trajectory in communicative AI. As shown in previous chapters, computer interface design emerged as a form of collaboration in which users do not so much "fall" into deception as they participate in constructing the representation that creates the very capacity for their interaction with computing systems. In line with this mechanism, there is a structural ambivalence in AI assistants that results from the complex exchanges between the software and the user, whereby the machine is adapted to the human so that the human can project its own meanings into the machine.

Voice assistants such as Alexa and Siri are not trying to fool anyone into believing that they are human. Yet, as I have shown, their functioning is strictly bounded to a banal form of deception that benefits from cutting-edge technical innovations in neural networks as well as from dramaturgical strategies established throughout decades of experimentation within communicative AI. Despite not being credible as humans, therefore, voice assistants are still capable of fooling us. This seems a contradiction only so long as one believes that deception involves a binary decision: if we are, in other words, either "fooled" or "not fooled." Both the history of AI and the longer history of media demonstrate that this is not the case—that technologies incorporate deception in more nuanced and oblique ways than is usually acknowledged.

In the future, the dynamics of projection and representation embedded in the design of voice assistants might be employed as means of manipulation and persuasion. As Judith Donath perceptively notes, it is likely that AI assistants and virtual companions in the future will be monetized through advertisement, just as advertisement has been a crucial revenue for other media of mass consumptions, from television and newspapers to web search engines and social media.[83] This raises the question of how feelings such as empathy, aroused by AI through banal deception, might be manipulated in order to convince users to buy certain products or, even more worryingly, to vote for a given party or candidate. Moreover, as AI voice assistants are also interfaces through which people access Internet-based resources, we need to ask how banal deception can inform access to information on the web—especially considering that Alexa, Siri, and the like are all proprietary technologies that privilege the cloud-based services and priorities of their respective companies.

Rather than being restricted to the case of voice assistants, some of the dynamics examined in this chapter extend to a broad range of AI-based technologies. Beside "flagship" conversational AI such as Siri, Alexa and Google Assistant, a proliferation of virtual assistants, chatbots, social media bots, email bots and AI companions envisioned to work in more or less specific domains and to carry out diverse task have emerged in the last few years.[84] While only a few of them might be programmed to undertake forms of deliberate and straight-out deception in contexts such as phishing or disinformation campaigns, many incorporate in their design deceptive mechanisms similar to those highlighted in this chapter. The implications, as discussed in the conclusion, are broader and more complex than any single AI technology.

Conclusion

Our Sophisticated Selves

Since its inception, AI has been at the center of incessant controversies. Computer scientists, psychologists, philosophers, and publicists have argued, in different moments of its history, that the promise of achieving AI was an illusion if not straight-out trickery.[1] Others, by contrast, have strenuously defended the AI enterprise, arguing that what was not yet achieved in the present might occur in the future and pointing to the field's many practical achievements.[2] Is there an endpoint to this controversy? The problem, as Turing intuited in 1950, is that the meaning of the term *AI* can vary enormously. It can be used to describe software that performs tasks in very specific domains, as well as to envision the dream of a "general" or "strong" AI that equals humans in all kinds of activities and tasks. It can be interpreted as the mere simulation of intelligence or as the project to instill consciousness in machines.[3] As long as it remains so arduous to agree on its meaning, AI is likely to keep attracting controversy.

Often, the solution to an apparently irresolvable problem is to choose a different approach altogether. In this book, I have argued that the question whether AI is real or a deception is ill-posed since deception is in itself a central component of AI. Entering into dialogues with voice assistants or interacting with bots on social media, users will rarely confuse the machine with a human. This does not mean, however, that deception does not occur, as these tools mobilize mechanisms such as empathy, stereotyping, and previous interaction habits to shape our perceptions and uses of AI. I have proposed the concept of banal deception to refer to mechanisms of

Deceitful Media. Simone Natale, Oxford University Press (2021). © Oxford University Press.
DOI: 10.1093/oso/9780190080365.003.0008

deception that are subtler and more pervasive than the rigid opposition between authenticity and delusion suggests. The question, therefore, is not so much whether AI will reach consciousness or surpass human intelligence as how we can accommodate our relationships with technologies that rely not only on the power of computing but also on our liability to be deceived.

My approach is in continuity with recent attempts in communication and media studies to consider more closely how users understand computing and algorithms and how such understandings inform the outcome of their interactions with digital technologies.[4] I argue, in this regard, that the ways humans react to machines that are programmed to reproduce the appearance of intelligent behaviors represent a constitutive element of what is commonly called AI. Artificial intelligence technologies are not just designed to interact with human users: they are designed to fit specific characteristics of the ways users perceive and navigate the external world. Communicative AI becomes more effective not only by evolving from a technical standpoint but also by profiting from the social meanings humans project onto situations and things.

Throughout the book, I have taken special care to highlight that banal deception is not forcefully malicious, as it always contributes some form of value to the user. Yet the fact that banal deception improves the functionality of interactions with AI does not mean that it is devoid of problems and risks. On the contrary, precisely because they are so subtle and difficult to recognize, the effects of banal deception are deeper and wider than any forms of straight-out delusion. By appropriating the dynamics of banal deception, AI developers have the potential to affect the deeper structures of our social lives and experiences.

One crucial challenge has to do with the fact that banal deception already bears within it the germs of straight-out deception. I have shown, for instance, how the dynamics of projection and stereotyping make it easier for AI voice assistants to accommodate our existing habits and social conventions. Organizations and individuals can exploit these mechanisms for political and marketing purposes, drawing on the feeling of empathy that a humanlike assistant stimulates in consumers and voters.[5] Other mechanisms that underpin banal deception can be used maliciously or in support of the wrong agendas. Exposing the dynamics of banal deception, in this sense, provides a further way to expose the most troublesome and problematic appropriations of AI and to counteract the power that algorithms and data administered by public and private institutions hold over us.

In chapter 6, I undertook a critical analysis of the banal deception mechanisms embedded in AI voice assistants. The same analytical work could be undertaken with regard to other AI technologies and systems.[6] For what concerns robotics, for instance, the question is to what extent the dynamics of banal deception that characterize AI voice assistants will be incorporated into robots that bear a physical resemblance to humans.[7] In contrast with the low definition of AI assistants, which employ verbal or written communication to interact with humans, the robots of the future may develop into a high-definition, "hot" medium that dissolves the distinction between banal and deliberate deception. To make another example, concerning deepfakes (i.e., AI-based technologies that produce videos or images in which the face of a person is replaced with someone else's likeness, so that, for instance, famous persons or politicians can appear to be pronouncing words that they have never said), the "banal" impression of reality that technologies of moving images create is what makes their deliberate deception so effective.[8]

An important phenomenon this book has touched on only marginally is the proliferation of bots on social media. While this has recently been the subject of much attention, discussions have mostly focused on what happens when bots pretend to be human.[9] Less emphasis has been given to the fact that deception does not only occur when bots are substituted for humans. An interesting example, in this regard, is the chatbot impersonating Israeli prime minister Benjamin Netanyahu that was used on Facebook in the 2019 elections. Although it was made quite clear to users that the chatbot was not Netanyahu himself, it was still useful to create a sense of proximity to the candidate as well as to collect information about potential voters to be targeted through other media.[10]

Even when no deliberate deception is involved, the development of communicative AI based on banal deception will spark unprecedented changes in our relationships with machines and, more broadly, in our social lives. Continuous interactions with Alexa, Siri, and other intelligent assistants do not only make users more ready to accept that AI is taking up an increasing number of tasks. These interactions also impact on the capacity to distinguish exchanges that offer just an appearance of sociality from interactions that actually include the possibility of empathy on the part of our interlocutor.[11] Therefore, as banal deception mechanisms disappear into the fabric of our everyday lives, it will become increasingly difficult to maintain clear distinctions between machines and humans at the social and cultural levels. Artificial intelligence's influence on social habits and behaviors, moreover, will also involve situations where no machines are involved. There are growing concerns, for instance, that stereotypical constructions of gender, race, and class embedded in communicative AI stimulate people to reproduce the same prejudices in other contexts.[12] Likewise, interactions between children and AI may shape their relationships not only with computers but also with humans, redefining the dynamics of social contacts within and outside the family.[13]

The increasing development of deep learning algorithms and their relationship to banal deception also calls for further scrutiny. As I have shown, banal deception emerges from specific bodies of knowledge about humans, against which the medium is adapted to a particular image of prospective users. Building more credible chatbots and social media bots, for instance, requires knowledge concerning human conversational behavior; designing more effective voice assistants draws from knowledge of how people react to different kinds of voices.[14] The collection of knowledge about users, however, is never a neutral process. The modeling of the "human" against which AI's banal deception emerged excluded differences of gender, race, and class, resulting in forms of bias and inequality that have been incorporated in the design of computing technologies up to the present day. In this context, deep learning is not only significant because it advances technologies such as speech processing, simulating verbal behavior more effectively than ever. Deep learning is also significant because with it, the modeling of the user is achieved autonomously by the neural networks. Large masses of data about users' behaviors are harvested and employed to "train" the system so that it can carry out complex tasks.[15] The question therefore arises: to what extent can or should the modeling of the human that underpins banal deception be devolved to the agency of autonomous machines?

To be fair, the fact that the modeling of the human can be constructed algorithmically does not mean that developers do not hold any control over these systems. The functioning of deep learning, after all, strictly depends on the data that are fed to the system. The possibility of analyzing and correcting the bias of these systems is therefore always open. Supervising neural networks, however, is expensive and difficult at a technical level. As data have become a form of commodity with significant value, moreover, companies and even public institutions might not be ready to renounce the potential economic benefit that using such data entails. Deep learning, in this sense, will only make the need to consider the impact of banal deception—and of the modeling of the human that it entails—more urgent and important.

In order not to be found unprepared for these and other challenges, computer scientists, software developers, and designers have to take the potential outcomes of banal deception seriously into account. These experts should develop a range of instruments to ensure that deception is restricted to applications that are genuinely functional and useful for users.

To quote Joseph Weizenbaum, the limits for the applicability of AI "cannot be settled by asking questions beginning with 'can'" but should instead be posed only in terms of "oughts."[16] Developers of AI need to seriously reflect on the question of deception. The call to develop stricter professional standards for the AI community to prevent the production of deceptively "human" systems is not new, but a broader debate on the issue of deception needs to take place.[17] Computer scientists and software designers are usually hesitant to employ the term *deception* in reference to their work. Yet not only the less subtle, deliberate forms of deception need to be acknowledged and addressed as such. Acknowledging the incorporation of banal deception into AI calls for an ethics of fairness and transparency between the vendor and the user that should not only focus on potential misuses of the technology but on interrogating the outcomes of different design features and mechanisms embedded in AI. The AI community also needs to consider the principles of transparent design and user-friendly interfaces, developing novel ways to make deception apparent and helping users to better navigate the barriers between banal and deliberate deception. The fact that technologies such as speech processing and natural language generation are becoming more easily available to individuals and groups makes these endeavors all the more pressing.

One complication, in this regard, is the difficulty of attributing agency and responsibility to developers of "intelligent" systems. As David Gunkel stresses, "the assignment of culpability is not as simple as it might first appear to be," especially because we are dealing with technologies that stimulate users to attribute agency and personality to them.[18] Nevertheless, software is always constructed with action in mind, and programmers always aim to produce a certain outcome.[19] Although it is difficult to establish what the creators of software have had in mind, elements of this can be reconstructed retrospectively, moving from the technical characteristics of the technology and the economic and institutional context of its production—as I have endeavored to do through the critical analysis of voice assistants.

When undertaking such critical work, we should remember that any representation of the user—the "human" around whom developers and companies construct communicative AI technologies—is itself the fruit of a cultural assessment affected by its own bias and ideology. For example, a company's decision to give a feminine name and voice as default to its voice assistant might be based on research about people's perceptions of female and male voices, but this research remains embedded in specific methodological regimes and cultural frameworks.[20] Finally, we should not disregard the fact that users themselves have agency, which may subvert and reframe any expectation or prevision about the outcomes of human-machine interactions. Emphasizing the role of programmers and companies that produce and disseminate AI technologies should not lead to denying the active role of users. On the contrary, banal deception calls for the user to actively engage with the deceptive mechanisms, which makes the user an active, crucial component in the functioning of every interactive AI system.

Ultimately, what AI calls into question is the very essence of who we are. But not so much because it makes us lose sight of what it means to be human. Rather, the key message of AI is that our vulnerability to deception is part of what defines us. Humans have a distinct capacity to project intention, intelligence, and emotions onto others. This is as much a burden as a resource: after all, this is what makes us capable of entertaining meaningful social interactions with others. But it also makes us prone to be deceived by nonhuman interlocutors that simulate intention, intelligence, and emotions.

The secret to navigate banal deception, in this regard, lies in our capacity to maintain a skeptical stance in interactions with computing and digital media. Refusing to engage with AI altogether is unfortunately not a real option: even if one disconnects from all devices, AI will continue to inform one's social life indirectly through the impact it has on others.[21] What we can seek, however, is a balance between our capacity on the one side to profit from the tools AI places at our disposal and our ability on the other side to adopt an informed and reflective attitude toward AI. In order to achieve this, we must resist the normalization of the deceptive mechanisms embedded in contemporary AI and the silent power that digital media companies exercise over us. We should never cease to interrogate how the technology works, even while we are trying to accommodate it in the fabric of everyday life. As Margaret Boden has pointed out, the outcome of any form of deception depends after all on the sophistication of the user.[22] The most sophisticated technologies, therefore, demand of us that we follow through, becoming more and more sophisticated ourselves.

NOTES

INTRODUCTION

1. "Google's AI Assistant Can Now Make Real Phone Calls," 2018, available at https://www.youtube.com/watch?v=JvbHu_bVa_g&time_continue=1&app=desktop (retrieved 12 January 2020). See also O'Leary, "Google's Duplex."
2. As critical media scholar Zeynep Tufekci put it in a tweet that circulated widely. The full thread on Twitter is readable at https://twitter.com/zeynep/status/994233568359575552 (retrieved 16 January 2020).
3. Joel Hruska, "Did Google Fake Its Duplex AI Demo?," *ExtremeTech*, 18 March 2018, available at https://www.extremetech.com/computing/269497-did-google-fake-its-google-duplex-ai-demo (retrieved 16 January 2020).
4. See, among many others, on the one side, Minsky, "Artificial Intelligence"; Kurzweil, *The Singularity Is Near*; and on the other side, Dreyfus, *What Computers Can't Do*; Smith, *The AI Delusion*.
5. Smith and Marx, *Does Technology Drive History?*; Williams, *Television*; Jones, "The Technology Is Not the Cultural Form?"
6. Goodfellow, Bengio, and Courville, *Deep Learning*.
7. Benghozi and Chevalier, "The Present Vision of AI . . . or the HAL Syndrome."
8. See chapter 2.
9. Gombrich, *Art and Illusion*.
10. There are, however, some relatively isolated but still significant exceptions. Adar, Tan, and Teevan, for instance, distinguish between malicious and benevolent deception, which they describe as "deception aimed at benefitting the user as well as the developer." Such a benevolent form of deception, they note, "is ubiquitous in real-world system designs, although it is rarely described in such terms." Adar et al., "Benevolent Deception in Human Computer Interaction," 1. Similarly, Chakraborti and Kambhampati observe that the obvious outcome of embedding models of mental states of human users into AI programs is that it opens up the possibility of manipulation. Chakraborti and Kambhampati, "Algorithms for the Greater Good!" From a different perspective, Nake and Grabowski have conceptualized communications between human and machine as "pseudo-communication," arguing for the importance of a semiotic perspective to understand human-computer interaction. Nake and Grabowski, "Human–Computer Interaction Viewed as Pseudo-communication." See also Castelfranchi and Tan, *Trust and Deception in Virtual Societies*; Danaher, "Robot Betrayal."

11. Coeckelbergh, "How to Describe and Evaluate 'Deception' Phenomena"; Schuetzler, Grimes, and Giboney, "The Effect of Conversational Agent Skill on User Behavior during Deception"; Tognazzini, "Principles, Techniques, and Ethics of Stage Magic and Their Application to Human Interface Design."
12. DePaulo et al., "Lying in Everyday Life"; Steinel and De Dreu, "Social Motives and Strategic Misrepresentation in Social Decision Making"; Solomon, "Self, Deception, and Self-Deception in Philosophy"; Barnes, *Seeing through Self-Deception*. A broad, traditional definition of deception is "to cause to believe what is false"; see Mahon, "The Definition of Lying and Deception."
13. Acland, *Swift Viewing*.
14. Martin, *The Philosophy of Deception*, 3; Rutschmann and Wiegmann, "No Need for an Intention to Deceive?"
15. Wrathall, *Heidegger and Unconcealment*, 60.
16. Hoffman, *The Case against Reality*.
17. Pettit, *The Science of Deception*.
18. Hyman, *The Psychology of Deception*. Notable examples of early studies on deception from a psychology perspective include Triplett, "The Psychology of Conjuring Deceptions"; Jastrow, *Fact and Fable in Psychology*.
19. Parisi, *Archaeologies of Touch*; Littlefield, *The Lying Brain*; Alovisio, "Lo schermo di Zeusi"; Sterne, *The Audible Past*.
20. Martin, *The Philosophy of Deception*, 3.
21. Caudwell and Lacey, "What Do Home Robots Want?"
22. The attribute "banal" has been applied in the social sciences in regard to other topics, more famously by Michael Billig with his insightful theorization of "banal nationalism." Although there are some points of analogy with the concept of banal deception, especially regarding its everyday character and the fact that banal nationalism is often overlooked and not interrogated as such, banal nationalism has been on the whole a useful source of inspiration rather than a direct reference to develop my theory. See Billig, *Banal Nationalism*; as well as Hjarvard, "The Mediatisation of Religion," which proposes the notion of "banal religion" in specific reference to Billig's work.
23. See, among others, Guzman, "Imagining the Voice in the Machine"; Reeves and Nass, *The Media Equation*; Turkle, *The Second Self*.
24. Guzman, "Beyond Extraordinary."
25. Ekbia, *Artificial Dreams*; Finn, *What Algorithms Want*.
26. Guzman, "Beyond Extraordinary," 84.
27. Papacharissi, *A Networked Self and Human Augmentics, Artificial Intelligence, Sentience*.
28. Chun, "On 'Sourcery,' or Code as Fetish."
29. Black, "Usable and Useful."
30. Porcheron et al., "Voice Interfaces in Everyday Life"; Guzman, *Imagining the Voice in the Machine*.
31. Nass and Moon, "Machines and Mindlessness"; Kim and Sundar, "Anthropomorphism of Computers."
32. Langer, "Matters of Mind," 289.
33. Guzman, "Making AI Safe for Humans."
34. Black, "Usable and Useful."
35. Ortoleva, *Miti a bassa intensità*. Marshall McLuhan employs the notion of "cool media" to describe such media. McLuhan, *Understanding Media*.
36. Guzman, *Imagining the Voice in the Machine*.

37. Hepp, "Artificial Companions, Social Bots and Work Bots." On discussions of perceptions of robots, see the concept of the "uncanny valley," originally proposed in Mori, "The Uncanny Valley" (the original text in Japanese was published in 1970). It is also worth noting that AI voice assistants are "disembodied" only to the extent that they are not given a proper physical "body" whose movements they control, as in robots; however, all software has to some extent its own materiality, and AI voice assistants in particular are always embedded in material artifacts such as smartphones or smart speakers. See, on the materiality of software, Kirschenbaum, *Mechanisms*, and specifically on AI voice assistants, Guzman, "Voices in and of the Machine."
38. Akrich, "The De-scription of Technical Objects." See also Feenberg, *Transforming Technology*; Forsythe, *Studying Those Who Study Us*.
39. Chakraborti and Kambhampati, "Algorithms for the Greater Good!"
40. Interestingly, deception plays a key role not only in how knowledge about human psychology and perception is used but also in how it is collected and accumulated. Forms of deception, in fact, have played a key role in the designing of experimental psychological studies so as to mislead participants in such a way that they remain unaware of a study's actual purposes. See Korn, *Illusions of Reality*.
41. Balbi and Magaudda, *A History of Digital Media*.
42. Bucher, *If . . . Then*, 68.
43. Towns, "Towards a Black Media Philosophy."
44. Sweeney, "Digital Assistants."
45. Bourdieu, *Outline of a Theory of Practice*.
46. Guzman, "The Messages of Mute Machines."
47. For a clear and compelling description of the remits of "communicative AI," see Guzman and Lewis, "Artificial Intelligence and Communication."
48. Hepp, "Artificial Companions, Social Bots and Work Bots."
49. Guzman, *Human-Machine Communication*; Gunkel, "Communication and Artificial Intelligence"; Guzman and Lewis, "Artificial Intelligence and Communication." For a discussion of the concept of medium in AI and communication studies, see Natale, "Communicating with and Communicating Through."
50. Doane, *The Emergence of Cinematic Time*; Hugo Münsterberg, *The Film: A Psychological Study*.
51. Sterne, *The Audible Past*; Sterne, *MP3*. Even prominent writers, such as Edgar Allan Poe in his "Philosophy of Composition," have examined the capacity of literature to achieve psychological effects through specific stylistic means. Poe, *The Raven; with, The Philosophy of Composition*.
52. As shown, for instance, by the controversy about Google Duplex mentioned at the beginning of this introduction.
53. Bottomore, "The Panicking Audience?"; Martin Loiperdinger, "Lumière's Arrival of the Train"; Sirois-Trahan, "Mythes et limites du train-qui-fonce-sur-les-spectateurs."
54. Pooley and Socolow, "War of the Words"; Heyer, "America under Attack I"; Hayes and Battles, "Exchange and Interconnection in US Network Radio."
55. There are, however, important exceptions; among works of media history that make important contributions to the study of deception and media, it is worth mentioning Sconce, *The Technical Delusion*; Acland, *Swift Viewing*.
56. See, for instance, McCorduck, *Machines Who Think*; Boden, *Mind as Machine*.

57. Riskin, "The Defecating Duck, or, the Ambiguous Origins of Artificial Life"; Sussman, "Performing the Intelligent Machine"; Cook, *The Arts of Deception.*
58. Geoghegan, "Visionäre Informatik." Sussman, "Performing the Intelligent Machine."
59. Gitelman, *Always Already New*; Huhtamo, *Illusions in Motion*; Parikka, *What Is Media Archaeology?*
60. This is why predictions about technology so often fail. See on this Ithiel De Sola Pool et al., "Foresight and Hindsight"; Natale, "Introduction: New Media and the Imagination of the Future."
61. Park, Jankowski, and Jones, *The Long History of New Media*; Balbi and Magaudda, *A History of Digital Media.*
62. Suchman, *Human-Machine Reconfigurations.*
63. For an influential example, see Licklider and Taylor, "The Computer as a Communication Device."
64. Appadurai, *The Social Life of Things*; Gell, *Art and Agency*; Latour, *The Pasteurization of France.*
65. Edwards, "Material Beings."
66. Reeves and Nass, *The Media Equation.*
67. See, among many others, Nass and Brave, *Wired for Speech*; Nass and Moon, "Machines and Mindlessness."
68. Turkle's most representative works are *Reclaiming Conversation*; *Alone Together*; *Evocative Objects*; *The Second Self*; *Life on the Screen.*

CHAPTER 1

1. Cited in Monroe, *Laboratories of Faith*, 86. See also Natale, *Supernatural Entertainments.*
2. Noakes, "Telegraphy Is an Occult Art."
3. "Faraday on Table-Moving."
4. Crevier, *AI*; McCorduck, *Machines Who Think*; Nilsson, *The Quest for Artificial Intelligence.*
5. Martin, "The Myth of the Awesome Thinking Machine"; Ekbia, *Artificial Dreams*; Boden, *Mind as Machine.*
6. Shieber, *The Turing Test*; Saygin, Cicekli, and Akman, "Turing Test."
7. Crevier, *AI.*
8. Some of these thinkers' biographies are a great access point to their lives, contributions, and thinking, as well as to the early history of AI and computing. See especially Soni and Goodman, *A Mind at Play*; Conway and Siegelman, *Dark Hero of the Information Age*; Copeland, *Turing.*
9. McCulloch and Pitts, "A Logical Calculus of the Ideas Immanent in Nervous Activity."
10. Wiener, *Cybernetics*. Wiener nonetheless believed that digital operations could not account for all chemical and organic processes that took place in the human organism, and advocated the inclusion of processes of an analog nature to study and replicate the human mind.
11. Nagel, "What Is It like to Be a Bat?" Philosopher John Searle presented an influential dismissal of the Turing test based on a similar perspective. In an article published in 1980, Searle imagined a thought experiment called the Chinese Room, in which a computer program could be written to participate in a conversation in Chinese so convincingly that a Chinese native speaker would exchange it for a human being. He reasoned that a person who did not speak

Chinese would be able to do the same by sitting in the "Chinese room" and following the same code of instructions through which the program operated. Yet, even if able to pass the Turing test, this person would not be able to understand Chinese but merely to simulate such understanding by providing the correct outputs on the basis of the code of rules. Searle considered this to be a demonstration that the computer would not properly "understand" Chinese either. He used the Chinese Room to counteract inflated claims about the possibility of building "thinking machines" that have characterized much of the rhetoric around AI up to the present day. He believed in an irremediable difference between human intelligence and what could be achieved by machines, which made imitation the only goal at which AI could reasonably aim. Searle, "Minds, Brains, and Programs." See also, on why it is impossible to know what happens in others' minds and the consequences of this fact for the debate on AI, Gunkel, *The Machine Question*.

12. The consequent emphasis on behavior, instead of what happens "inside the machine," was advocated not only by Turing but by other key figures of early AI, including Norbert Wiener, who reasoned that "now that certain analogies of behavior are being observed between the machine and the living organism, the problem as to whether the machine is alive or not is, for our purpose, semantic and we are at liberty to answer it one way or the other as best suits our convenience." Wiener, *The Human Use of Human Beings*, 32.
13. Turing, "Computing Machinery and Intelligence," 433.
14. Luger and Chakrabarti, "From Alan Turing to Modern AI."
15. Bolter, *Turing's Man*; on the Turing test, computing, and natural language, see Powers and Turk, *Machine Learning of Natural Language*, 255.
16. For an overview of the debate, including key critical texts and responses to Turing, see Shieber, *The Turing Test*. Some have wondered if Turing did so intentionally to make sure that the article would spark reactions and debate—so that many would be forced to consider and discuss the problem of "machine intelligence." Gandy, "Human versus Mechanical Intelligence."
17. See, for instance, Levesque, *Common Sense, the Turing Test, and the Quest for Real AI*.
18. Computer scientist and critical thinker Jaron Lanier has been among the most perceptive in this regard. As he points out, "what the test really tells us . . . even if it's not necessarily what Turing hoped it would say, is that machine intelligence can only be known in a relative sense, in the eyes of a human beholder." As a consequence, "you can't tell if a machine has gotten smarter or if you're just lowered your own standards of intelligence to such a degree that the machine seems smart." Lanier, *You Are Not a Gadget*, 32.
19. As Jennifer Rhee perceptively notes, in the Turing test "the onus of success or failure does not rest solely on the abilities of the machine, but is at least partially distributed between machine and human, if not located primarily in the human." Rhee, "Misidentification's Promise."
20. See, for example, Levesque, *Common Sense, the Turing Test, and the Quest for Real AI*.
21. Turing, "Computing Machinery and Intelligence," 441. As noted by Helena Granström and Bo Göranzon, "Turing does not say: at the end of the century technology will have advanced so much that machines actually will be able to think. He says: our understanding of human thinking will have shifted towards formalised information processing, to the extent that it will no longer be

possible to distinguish it from a mechanical process." Granström and Göranzon, "Turing's Man," 23.

22. This argument was also made in Weizenbaum, *Computer Power and Human Reason*.
23. Turkle, *Life on the Screen*, 24. Similar dynamics of change in cultural perceptions was noticed by Turkle in other studies that looked at children and adults' perceptions of computing machines. "For today's children, the boundary between people and machines is intact. But what they see across the boundary has changed dramatically. Now, children are comfortable with the idea that inanimate objects can both think and have a personality. But they no longer worry if the machine is alive. They know it's not" (83). For what concerns adults, Turkle's studies lead her to conclude that "today, the controversy about computers does not turn on their capacity for intelligence but on their capacity for life" (84). Margaret Boden also notes changes in the use of words to refer to computers: "people now speak about computers in psychological terms as a routine, everyday, matter." Boden, *Minds as Machines*, 1356.
24. See on this Lamont, *Extraordinary Beliefs*.
25. Peters, *Speaking into the Air*.
26. Turkle, *Alone Together*.
27. Ceruzzi, *A History of Modern Computing*.
28. Surely, the idea already lingered in a nascent computer culture. Supporters of cybernetics, for instance, posited humans and machines as part of a system of communication and feedback loops that anticipated in many ways developments that followed; see for instance Wiener, *The Human Use of Human Beings*. But few if any early writings on computing went on to imagine future interactions between computers and humans in such a vivid way as Turing's "Computing Machinery and Intelligence."
29. Garfinkel, *Architects of the Information Society*.
30. Gunkel, *Gaming the System*, 133.
31. Christian, *The Most Human Human*, 37. See also Bratton, "Outing Artificial Intelligence: Reckoning with Turing Test."
32. Enns, "Information Theory of the Soul."
33. Turing, "Computer Machinery and Intelligence," 434.
34. Gitelman, *Scripts, Grooves, and Writing Machines*; Kittler, *Gramophone, Film, Typewriter*.
35. By achieving anonymous, disembodied communication, the test also foregrounds mediated communication in the Internet. It is indeed natural for the contemporary reader to associate the design of the test with situations encountered online. In a chatroom, a web forum, or on social media, when conversation is limited to text and the identities of other users are conveyed through little more than a nickname and an image, one cannot be entirely sure if users are who or what they pretend to be—which fits quite well the objectives of the Turing test.
36. Copeland, *The Essential Turing*, 420.
37. McLuhan, *Understanding Media*.
38. Media archaeologist Erkki Huhtamo demonstrated that these visions were triggered by reflections on potential anthropomorphic changes as a consequence of technological innovation. Huhtamo, "Elephans Photographicus."
39. Copeland, *The Essential Turing*, 420.
40. Guzman, "Voices in and of the Machine."

41. "Turing himself was always careful to refer to 'the game.' The suggestion that it might be some sort of test involves an important extension of Turing's claims." Whitby, "The Turing Test," 54.
42. Broussard, for instance, recently pointed out that this was due to the fact that computer scientists "tend to like certain kinds of games and puzzles." Broussard, *Artificial Unintelligence*, 33. See also Harnad, "The Turing Test Is Not a Trick."
43. Johnson, *Wonderland*.
44. Soni and Goodman, *A Mind at Play*; Weizenbaum, "How to Make a Computer Appear Intelligent"; Samuel, "Some Studies in Machine Learning Using the Game of Checkers."
45. Newell, Shaw, and Simon, "Chess-Playing Programs and the Problem of Complexity." Before his proposal of the Imitation Game, Turing suggested chess as a potential test bed for AI. Turing, "Lecture on the Automatic Computing Engine," 394; Copeland, *The Essential Turing*, 431.
46. Ensmensger, "Is Chess the Drosophila of Artificial Intelligence?"
47. Kohler, *Lords of the Fly*.
48. Rasskin-Gutman, *Chess Metaphors*; Dennett, "Intentional Systems."
49. Ensmenger, "Is Chess the Drosophila of Artificial Intelligence?," 7. See also Burian, "How the Choice of Experimental Organism Matters: Epistemological Reflections on an Aspect of Biological Practice."
50. Franchi, "Chess, Games, and Flies"; Ensmensger, "Is Chess the Drosophila of Artificial Intelligence?"; Bory, "Deep New."
51. As Galloway puts it, "without action, games remain only in the pages of an abstract rule book"; Galloway, *Gaming*, 2. See also Fassone, *Every Game Is an Island*. In addition, one of the principles of game theory is that, even if the players' actions can be formalized as a mathematical theory of effective strategies, only the actual play of a game activates the abstract formalization of a game. Juul, *Half-Real*.
52. Friedman, "Making Sense of Software."
53. Laurel, *Computers as Theatre*, 1.
54. Galloway, *Gaming*, 5. A criticism of this argument points to the problem of intentionality and the frame through which specific situations and actions are understood and interpreted. While at the level of observed behavior computer and human players may look the same, only the latter may have the capacity to distinguish the different frames of playful and not playful activities, e.g., the playing of a war game from war itself. One might however respond to this criticism with the same behavioral stance exemplified by the Turing test, which underscores the problem of what happens inside the machine to focus instead on observable behavior. See Bateson, "A Theory of Play and Fantasy"; Goffman, *Frame Analysis*.
55. Galloway, *Gaming*.
56. Dourish, *Where the Action Is*.
57. Christian, *The Most Human Human*, 175.
58. See, for instance, Block, "The Computer Model of the Mind."
59. Korn, *Illusions of Reality*.
60. See chapter 5 on the Loebner Prize contest, organized yearly since the early 1990s, in which computer programmers enter their chatbots in the hope of passing a version of the Turing test.
61. Geoghegan, "Agents of History." Similarly, Lanier observes that both chess and computers originated as tools of war: "the drive to compete is palpable in both

computer science and chess, and when they are brought together, adrenaline flows." Lanier, *You Are not a Gadget*, 33.

62. Copeland, "Colossus."
63. Whitby, "The Turing Test."
64. Goode, "Life, but Not as We Know It."
65. Huizinga, *Homo Ludens*, 13.
66. This resonates in contemporary experiences with systems such as Siri and Alexa, where human users engage in jokes and playful interactions, questioning the limits and the potential of AI's imitation of humanity. See Andrea L. Guzman, "Making AI Safe for Humans."
67. Scholars have suggested that this was connected to an underlying concern of Turing with definitions of gender. See, among others, Shieber, *The Turing Test*, 103; Bratton, "Outing Artificial Intelligence." Representations of gender and chatbots playing a version of the Turing test are further discussed in chapter 5.
68. Hofer, *The Games We Played*.
69. Dumont, *The Lady's Oracle*.
70. Ellis, *Lucifer Ascending*, 180.
71. See, on this, Natale, *Supernatural Entertainments*, especially chapter 2. Anthony Enns recently argued that the Turing test "resembled a spiritualist séance, in which sitters similarly attempted to determine whether an invisible spirit was human by asking a series of questions. Like the spirits that allegedly manifested during these séances, an artificial intelligence was also described as a disembodied entity that exists on an immaterial plane that can only be accessed through media technologies. And just as séance-goers often had difficulty determining whether the messages received from spirits could be considered evidence of a genuine presence, so too did scientists and engineers often have difficulty determining whether the responses received from machines could be considered evidence of genuine intelligence. These two phenomena were structurally similar, in other words, because they posited that there was no essential difference between identity and information or between real intelligence and simulated intelligence, which effectively made humans indistinguishable from machines." Enns, "Information Theory of the Soul."
72. Lamont, *Extraordinary Beliefs*.
73. As American show business pioneer P. T. Barnum reportedly observed, "the public appears disposed to be amused even when they are conscious of being deceived." Cook, *The Arts of Deception*, 16.
74. See, for instance, Von Hippel and Trivers, "The Evolution and Psychology of Self Deception."
75. Marenko and Van Allen, "Animistic Design."
76. McLuhan, *Understanding Media*.
77. Guzman and Lewis, "Artificial Intelligence and Communication."
78. See Leja, *Looking Askance*; Doane, *The Emergence of Cinematic Time*; Sterne, *MP3*.

CHAPTER 2

1. Kline, "Cybernetics, Automata Studies, and the Dartmouth Conference on Artificial Intelligence"; Hayles, *How We Became Posthuman*.
2. See, for instance, Crevier, *AI*.
3. Solomon, *Disappearing Tricks*; Gunning, "An Aesthetic of Astonishment." Practicing theatrical conjurers such as Georges Méliès were among the first motion picture exhibitionists and filmmakers, and many early spectators

experienced cinema as part of magic stage shows. See, among others, Barnouw, *The Magician and the Cinema*; North, "Magic and Illusion in Early Cinema"; Leeder, *The Modern Supernatural and the Beginnings of Cinema*.

4. During, *Modern Enchantments*.
5. Chun, *Programmed Visions*. It is for this reason that there is an element of creepiness and wonder in how algorithms inform everyday life—so that sometimes, when one sees one's tastes or searches anticipated by digital platforms, one cannot help but wonder if Amazon or Google are reading one's mind. See, on this, Natale and Pasulka, *Believing in Bits*.
6. Martin, "The Myth of the Awesome Thinking Machine," 122. Public excitement about AI was also fueled by science fiction films and literature. On the relationship between the AI imaginary and fictional literature and film, see Sobchack, "Science Fiction Film and the Technological Imagination"; Goode, "Life, but Not as We Know It"; Bory and Bory, "I nuovi immaginari dell'intelligenza artificiale."
7. Martin, "The Myth of the Awesome Thinking Machine," 129.
8. As historian of AI Hamid Ekbia points out, "what makes AI distinct from other disciplines is that its practitioners 'translate' terms and concepts from one domain into another in a systematic way." Ekbia, *Artificial Dreams*, 5. See also Haken, Karlqvist, and Svedin, *The Machine as Metaphor and Tool*.
9. Minsky, *Semantic Information Processing*, 193.
10. Minsky, *Semantic Information Processing*.
11. On the role of analogies and metaphors in scientific discourse, see Bartha, "Analogy and Analogical Reasoning." The comparison between artificial and biological life could go so far as to include elements of humanity that surpassed the boundaries of mere rational thinking, to include feelings and emotions. In 1971, for instance, an article in *New Scientist* was titled "Japanese Robot Has Real Feeling." By reading the article with more attention, one could understand that the matter of the experiments was not so much human emotions as the capacity of a robot to simulate tactile perception by gaining information about an object through contact. Playing with the semantic ambiguity of the words "feeling" and "feelings," and alluding to human emotions well beyond basic tactile stimuli, the author added a considerable amount of sensationalism to his report. Anonymous, "Japanese Robot Has Real Feeling," 90. See on this Natale and Ballatore, "Imagining the Thinking Machine."
12. Hubert L. Dreyfus, *What Computers Can't Do*, 50–51.
13. Philosopher John Haugeland labeled this approach "Good Old-Fashioned Artificial Intelligence" (GOFAI), to distinguish it from more recent sub-symbolic, connectionist, and statistical techniques. Haugeland, *Artificial Intelligence*.
14. The concept of the "black box," which describes technologies that provide little or no information about their internal functioning, can therefore apply to computers as well as to the human brain.
15. Cited in Smith, *The AI Delusion*, 23. His words echoed Weizenbaum's worried remarks about the responsibility of computer scientists to convey accurate representations of the functioning and power of computing systems. Weizenbaum, *Computer Power and Human Reason*.
16. For an in-depth discussion, see Natale and Ballatore, "Imagining the Thinking Machine."
17. Minsky, "Artificial Intelligence," 246. On Minsky's techno-chauvinism, see Broussard, *Artificial Unintelligence*.

18. In a frequently cited article of the time, Armer set up the goal of analyzing attitudes toward AI with the hope to "improve the climate which surrounds research in the field of machine or artificial intelligence." Armer, "Attitudes toward Intelligent Machines," 389.
19. McCorduck, *Machines Who Think*, 148.
20. Messeri and Vertesi, "The Greatest Missions Never Flown." The case of Moore's law is a good example in the field of computer science of the ways projections about future accomplishments may motivate specific research communities to direct their expectations and work toward the achievement of certain standards, but also to maintain their efforts within defined boundaries. See Brock, *Understanding Moore's Law*.
21. Notable examples include Kurzweil, *The Singularity Is Near*; Bostrom, *Superintelligence*. For what concerns fiction, examples are numerous and include films such as the different installments of the *Terminator* saga, *Her* (2013), and *Ex Machina* (2014).
22. Broussard, *Artificial Unintelligence*, 71. See also Stork, *HAL's Legacy*.
23. McLuhan, *Understanding Media*.
24. Doane, *The Emergence of Cinematic Time*; Münsterberg, *The Film*.
25. Sterne, *The Audible Past*; Sterne, *MP3*.
26. In his overview of the history of human-computer interaction design up to the 1980s, Grudin argues that "the location of the 'user interface' has been pushed farther and farther out from the computer itself, deeper into the user and the work environment." The focus moved from directly manipulating hardware in early mainframe computers to different levels of abstraction through programming languages and software and to video terminals and other devices that adapted to the perceptive abilities of users, wider attempts to match cognitive functions, and finally approaches to moving the interface into wider social environments and groups of users. Grudin, "The Computer Reaches Out," 246. For a history of human-computer interaction that also acknowledges that this field has a history before the computer, see Mindell, *Between Human and Machine*.
27. This applies especially to the initial decades of development in computer science, when human-computer interaction had not yet coalesced into an autonomous subfield. Yet the problems involved by human-computer interaction remained relevant to AI throughout its entire history, and even if it might be useful to separate the two areas in certain contexts and to tackle specific problems, it is impossible to understand the close relationship between AI and communication without considering AI and human-computer interaction along parallel lines. Grudin, "Turing Maturing"; Dourish, *Where the Action Is*.
28. Marvin Minsky, for instance, apologized at the end of a 1961 article for the fact that "we have discussed here only work concerned with more or less self-contained problem-solving programs. But as this is written, we are at last beginning to see vigorous activity in the direction of constructing usable time-sharing or multiprogramming computing systems. With these systems, it will at last become economical to match human beings in real time with really large machines. This means that we can work toward programming what will be, in effect, "thinking aids." In the years to come, we expect that these man-machine systems will share, and perhaps for a time be dominant, in our advance toward the development of 'artificial intelligence.'" Minsky, "Steps toward Artificial Intelligence," 28.

29. See, on this, Guzman, *Human-Machine Communication*; Guzman and Lewis, "Artificial Intelligence and Communication."
30. Licklider, "Man-Computer Symbiosis." On the impact of Licklider's article on the AI and computer science of the time, see, among others, Edwards, *Closed Worlds*, 266. Although Licklider's article brought the concept of symbiosis to the center of debates in computer science, like-minded approaches had been developed before in cybernetics and in earlier works that looked at computing machines not only as processors but also as sharers of information. See Bush, "As We May Think"; Hayle, *How We Became Posthuman*. For an example of approaches that looked at the relationship between computers and communication outside the scope of AI, as in ergonomics and computer-supported collaborative work, see Meadow, *Man-Machine Communication*.
31. Licklider, "Man-Computer Symbiosis," 4.
32. Hayles, *How We Became Posthuman*.
33. Shannon, "The Mathematical Theory of Communication"; Wiener, *Cybernetics*. Interestingly, understandings of communication as disembodied, which emerged already in the nineteenth century with the development of electronic media, also stimulated spiritual visions based on the idea that communication could be wrested from its physical nature; this linked with information theory, as shown by Enns, "Information Theory of the Soul." See also Carey, *Communication as Culture*; Peters, *Speaking into the Air*; Sconce, *Haunted Media*.
34. Hayles, *How We Became Posthuman*; McKelvey, *Internet Daemons*, 27. As mentioned, the key tenet of early AI was that rational thinking was a form of computation, which could be replicated through symbolic logic programmed into computers.
35. Suchman, *Human-Machine Reconfigurations*.
36. See, for instance, Carbonell, Elkind, and Nickerson, "On the Psychological Importance of Time in a Time Sharing System"; Nickerson, Elkind, and Carbonell, "Human Factors and the Design of Time Sharing Computer Systems."
37. At the beginning of the computer age, the large computer mainframes installed at universities and other institutions were able to manage only one process at a time. Programs were executed in batch, each of them occupying all of the computer's resources at a given time. This meant that researchers had to take turns to access the machines. Gradually, a new approach called time-sharing emerged. In contrast with previous systems, time-sharing allowed different users to access a computer at the same time. Giving users the impression that the computer was responding in real time, it opened the way for the development of new modalities of user-computer interaction and enabled wider access to computer systems.
38. Simon, "Reflections on Time Sharing from a User's Point of View," 44.
39. Greenberger, "The Two Sides of Time Sharing."
40. Greenberger, "The Two Sides of Time Sharing," 4. A researcher at MIT who worked on Project MAC, Greenberger noted that the subject of time-sharing can be bisected into separate issues/problems: the system and the user. This distinction "between system and user is reflected in the double-edged acronym of MIT's Project MAC: Multi-Access Computer refers to the physical tool or system, whereas Machine-Aided Cognition expresses the hopes of the user" (2).
41. Broussard, *Artificial Unintelligence*; Hicks, *Programmed Inequality*.

42. Weizenbaum, "How to Make a Computer Appear Intelligent," 24. Weizenbaum's own work on natural language processing, chatbots, and deception is discussed in chapter 3.
43. Minsky, "Some Methods," 6.
44. Minsky, "Some Methods," 6.
45. Minsky, "Problems of Formulation for Artificial Intelligence," 39.
46. In an article published in *Scientific American* in 1966, for instance, the MIT scientist did not refrain from expressing his optimism for the prospects of AI: "Once we have devised programs with the capacity for self-improvement a rapid evolutionary process will begin. As the machine improves both itself and its model of itself, we shall begin to see all the phenomena associated with the terms 'consciousness,' 'intuition' and 'intelligence' itself. It is hard to say how close we are to this threshold, but once it is crossed the world will not be the same." Minsky, "Artificial Intelligence," 260.
47. Pask, "A Discussion of Artificial Intelligence," 167. Italics in original.
48. Pask, "A Discussion of Artificial Intelligence," 153.
49. Geoghegan, "Agents of History." Despite, of course, no supernatural events being involved, Shannon chose to use a term from parapsychology to illustrate how his machine worked. The concept of mind reading had emerged as a major subject of investigation in parapsychology and psychical research at the end of the nineteenth century, referring to the possibility that human minds might be able to communicate through extrasensorial means. By employing the idea of the "mind reading machine," Shannon metaphorically described the capacity of calculators to elaborate statistical data in order to anticipate human behaviors, but at the same time he alluded to the realm of paranormal phenomena, which is associated with a wide range of imaginary and fantastic myths. See Natale, "Amazon Can Read Your Mind."
50. Musès, *Aspects of the Theory of Artificial Intelligence*. See also Barad, *Meeting the Universe Halfway*.
51. McCarthy, "Information," 72.
52. Minsky, "Some Methods"; Weizenbaum, *Computer Power and Human Reason*, 189.
53. Manon, "Seeing through Seeing Through"; Couldry, "Liveness, 'Reality,' and the Mediated Habitus from Television to the Mobile Phone."
54. Emerson, *Reading Writing Interfaces*.
55. Writing in the 1990s, computer scientist Branda Laurel compared for instance human-computer interaction to a conversation. She argued that time-sharing introduced a conversational quality to human-computer interaction, enabling the construction of a "common ground" between the machine and the human. Laurel, *Computer as Theater*, 4.
56. Greenberger, "The Two Sides of Time Sharing."
57. Ceruzzi, *A History of Modern Computing*, 154.
58. McCorduck, *Machines Who Think*, 216–17.
59. In *Gameworld Interfaces*, Kristine Jørgensen defines interfaces "as the interconnection between different spheres and, in human-computer interaction, as the part of the system that allows the user to interact with the computer." Jørgensen, *Gameworld Interfaces*, 3. See also Hookway, *Interface*, 14.
60. Emerson, *Reading Writing Interfaces*, x; Galloway, *The Interface Effect*.
61. Chun, *Programmed Visions*, 66.
62. Laurel, *Computers as Theatre*, 66–67.
63. Hookway, *Interface*, 14.

64. Oettinger, "The Uses of Computers," 162.
65. Oettinger, "The Uses of Computers," 164. Emphasis mine.
66. Black, "Usable and Useful."
67. In a similar way, in fact, a magician performs a trick by hiding its technical nature so that the source of the illusion is unknown to the audience. See Solomon, *Disappearing Tricks*.
68. Emerson, *Reading Writing Interfaces*, xi.
69. As Carlo Scolari points out, the transparency of the interface is the utopia of every contemporary interface design; however, the reality is different, as even the simplest example of interaction hides an intricate network of interpretative negotiations and cognitive processes. An interface, therefore, is never neutral or transparent. Scolari, *Las leyes de la interfaz*.
70. Geoghegan, "Visionäre Informatik."
71. See Collins, *Artifictional Intelligence*.

CHAPTER 3

1. Guzman, "Making AI Safe for Humans."
2. Leonard, *Bots*, 33–34.
3. Russell and Norvig, *Artificial Intelligence*.
4. McCorduck, *Machines Who Think*, 253.
5. Zdenek, "Rising Up from the MUD," 381.
6. Weizenbaum, "ELIZA." For detailed but accessible explanations of ELIZA's functioning, see Pruijt, "Social Interaction with Computers"; Wardrip-Fruin, *Expressive Processing*, 28–32.
7. Uttal, *Real-Time Computers*, 254–55.
8. Weizenbaum, "ELIZA."
9. Weizenbaum, "How to Make a Computer Appear Intelligent."
10. Crevier, *AI*, 133. Crevier reports that in conversations with Weizenbaum, Weizenbaum told him that he had started his career "as a kind of confidence man": "the program used a ridiculously simple strategy with no look ahead, but it could beat anyone who played at the same naive level. Since most people had never played the game before, that included just everybody. . . . In a way, that was a forerunner to my later ELIZA, to establish my status as a charlatan or con man. But the other side of the coin is that I freely stated it. The idea was to create a powerful illusion that the computer was intelligent. I went to considerate trouble in the paper to explain that there wasn't much behind the scenes, that the machine wasn't thinking. I explained the strategy well enough that anybody could write the program, which is the same thing I did with ELIZA" (133).
11. Weizenbaum, "How to Make a Computer Appear Intelligent," 24.
12. Weizenbaum, "Contextual Understanding by Computers." This element is implicit in Turing's proposal. As Edmonds notes, "the elegance of the Turing Test (TT) comes from the fact that it is not a requirement upon the mechanisms needed to implement intelligence but on the ability to fulfil a role. In the language of biology, Turing specified the niche that intelligence must be able to occupy rather than the anatomy of the organism. The role that Turing chose was a social role—whether humans could relate to it in a way that was sufficiently similar to a human intelligence that they could mistake the two." Edmonds, "The Constructibility of Artificial Intelligence (as Defined by the Turing Test)," 419.
13. Weizenbaum, "ELIZA," 37. The notion of script later became a common trope in technical descriptions of conversational software, which frequently represent

interactions between users and programs through conventions of a dramatic dialogue; see Zdenek, "Artificial Intelligence as a Discursive Practice," 353. Also, script theory in psychology posits that human behavior is informed by patterns (called scripts) that provide programs for action, and the notion has been employed in AI research to achieve the goal of replicating human behavior, as illustrated for instance in Schank and Abelson, *Scripts, Plans, Goals, and Understanding*. See also the use of the concept of script in Actor-Network Theory, especially Akrich, "The De-scription of Technical Objects."

14. In a conversation with AI historian Daniel Crevier, Weizenbaum told Crevier that "he had originally conceived of ELIZA as a barman, but later decided psychiatrists were more interesting." Crevier, *AI*, 136.
15. Weizenbaum, "ELIZA," 36–37.
16. Shaw, *Pygmalion*. For a reflection on the significance for AI of Cukor's *My Fair Lady* (1964), which was adapted from *Pygmalion*, see Mackinnon, "Artificial Stupidity and the End of Men."
17. Weizenbaum, *Islands in the Cyberstream*, 87.
18. Weizenbaum, "ELIZA," 36. See also Mackinnon, "Artificial Stupidity and the End of Men."
19. Weizenbaum, *Islands in the Cyberstream*, 88.
20. Weizenbaum, *Computer Power and Human Reason*, 188.
21. Weizenbaum, "Contextual Understanding by Computers," 475; see also Weizenbaum, *Islands in the Cyberstream*, 190. As noted by Lucy Suchman, ELIZA "exploited the maxim that shared premises can remain unspoken: that the less we say in conversation, the more what is said is assumed to be self-evident in its meaning and implications. . . . Conversely, the very fact that a comment is made without elaboration implies that such shared background assumptions exist. The more elaboration or justification is provided, the less the appearance of transparence or self-evidence. The less elaboration there is, the more the recipient will take it that the meaning of what is provided should be obvious." *Human-Machine Reconfigurations*, 48.
22. See, for instance, Murray, *Hamlet on the Holodeck*, as well as Christian, *The Most Human Human*.
23. Lakoff and Johnson, *Metaphors We Live By*.
24. Weizenbaum, *Computer Power and Human Reason*, 189.
25. Weizenbaum, *Computer Power and Human Reason*, 190.
26. Edgerton, *Shock of the Old*; Messeri and Vertesi, "The Greatest Missions Never Flown." See also Natale, "Unveiling the Biographies of Media."
27. Edgerton, *Shock of the Old*, 17–18.
28. Christian, *The Most Human Human*, 263–64; Bory, "Deep New."
29. Zachary, "Introduction," 17.
30. Boden, *Mind as Machine*, 743. See also Barr, "Natural Language Understanding."
31. Weizenbaum, "ELIZA." Luger and Chakrabarti stresses that "the construction of any human artefact includes an implicit epistemic stance." Luger and Chakrabarti, "From Alan Turing to Modern AI," 321.
32. Martin, "The Myth of the Awesome Thinking Machine."
33. Weizenbaum, "On the Impact of the Computer on Society."
34. Weizenbaum, *Islands in the Cyberstream*, 89.
35. Brewster, *Letters on Natural Magic*. This tradition is arguably quite relevant to the history of cybernetics and artificial intelligence, as shown for instance by the many playful devices Claude Shannon used to build in order to demonstrate

"programming tricks" or the "living prototypes" of Heinz von Foerster's Biological Computer Laboratory. See Soni and Goodman, *A Mind at Play*, 243–53; Müggenburg, "Lebende Prototypen und lebhafte Artefakte."

36. Weizenbaum, *Islands in the Cyberstream*, 89.
37. The gender dimension of this anecdote is not to be overlooked, as well as the gendered identity assigned to chatbots from ELIZA to Amazon's Alexa. See, on this, Zdenek, "Rising Up from the MUD."
38. Boden, *Mind as Machine*, 1352.
39. Turkle, *The Second Self*, 110.
40. Benton, *Literary Biography*; Kris and Kurz, *Legend, Myth, and Magic in the Image of the Artist*; Ortoleva, "Vite Geniali."
41. Sonnevend, *Stories without Borders*.
42. Wilner, Christopoulos, Alves, and Guimarães, "The Death of Steve Jobs," 430.
43. Spufford and Uglow, *Cultural Babbage*.
44. Martin, "The Myth of the Awesome Thinking Machine"; Bory and Bory, "I nuovi immaginari dell'intelligenza artificiale."
45. Crevier, *AI*.
46. Weizenbaum, "ELIZA," 36.
47. Wardrip-Fruin, *Expressive Processing*, 32.
48. Colby, Watt, and Gilbert, "A Computer Method of Psychotherapy," 148–52. PARRY was described as "ELIZA with attitude" and programmed to play the role of a paranoid. Its effectiveness was measured against the fact that it was often diagnosed by doctors unaware that they were dealing with a computer program; see Boden, *Minds and Machines*, 370. Some believe that PARRY was superior to ELIZA because "it has personality," as argued by Mauldin, "ChatterBots, TinyMuds, and the Turing Test," 16. Yet ELIZA also, despite its unwillingness to engage more significantly with the user's inputs, is instilled with a personality. It is more minimal also because Weizenbaum wanted to make a point, showing that even a very simple system could be perceived as "intelligent" by some users.
49. McCorduck, *Machines Who Think*, 313–15.
50. Weizenbaum, "On the Impact of the Computer on Society"; Weizenbaum, *Computer Power and Human Reason*, 268–70; Weizenbaum, *Islands in the Cyberstream*, 81.
51. Weizenbaum, *Computer Power and Human Reason*, 269–70.
52. Weizenbaum, Joseph, "The Tyranny of Survival: The Need for a Science of Limits," *New York Times*, 3 March 1974, 425.
53. See, for an interesting example, Wilford, "Computer Is Being Taught to Understand English."
54. Weizenbaum, "Contextual Understanding by Computers," 479; see also Geoghegan, "Agents of History," 409; Suchman, *Human-Machine Reconfigurations*, 49.
55. Weizenbaum, *Computer Power and Human Reason*.
56. Weizenbaum, "Letters: Computer Capabilities," 201.
57. Davy, "The Man in the Belly of the Beast," 22.
58. Johnston employs the concept of "computational assemblage" to argue that "every computational machine is conceived of as a material assemblage (a physical device) conjoined with a unique discourse that explains and justifies the machine's operation and purpose. More simply, a computational assemblage is comprised of both a machine and its associated discourse, which together

determine how and why this machine does what it does." Johnston, *The Allure of Machinic Life*, x.

59. Weizenbaum, *Computer Power and Human Reason*, 269.
60. McCorduck, *Machines Who Think*, 309.
61. Weizenbaum, *Computer Power and Human Reason*, 157. Luke Goode has recently argued, entering indirectly into controversy with Weizenbaum, that metaphors and narratives (in particular, fiction) can indeed be useful to improve public understandings of AI and thus the governance of these technologies. Goode, "Life, but Not as We Know It."
62. Hu, *A Prehistory of the Cloud*; Peters, *The Marvelous Cloud*.
63. Kocurek, "The Agony and the Exidy"; Miller, "Grove Street Grimm.'"
64. See, for instance, "The 24 Funniest Siri Answers That You Can Test with Your iPhone."
65. Murray, *Hamlet on the Holodeck*, 68, 72.
66. Wardrip-Fruin, *Expressive Processing*, 24.
67. All the ELIZA avatars I found use Javascript. Yet it is in principle possible to provide a reliable copy of ELIZA, due to the repeatability of code. This would not be the case with contemporary AI systems such as Siri and Alexa, which employ machine learning techniques to constantly improve their performances and are therefore never the same. In this, they are similar after all to humans, whose physiology and psychology change constantly, making them in a certain sense "different" at any different moment in time. Jeff Shrager's website provides a fairly comprehensive genealogy of ELIZA reconstructions and scholarship; see Shrager, "The Genealogy of Eliza."
68. Boden, *Minds and Machines*, 1353.
69. Marino, "I, Chatbot," 8.
70. Turner, *From Counterculture to Cyberculture*; Flichy, *The Internet Imaginaire*; Streeter, *The Net Effect*.
71. King, "Anthropomorphic Agents," 5.
72. Wardrip-Fruin, *Expressive Processing*, 32.
73. Interestingly, Wardrip-Fruin also notes that the effect conjured in this way by ELIZA was broken down as the interaction continued, the limitations of the chatbot became evident, and therefore the user's understanding of the internal processes at work improved. "In this context, it is interesting to note that most systems of control that are meant to appear intelligent have extremely restricted methods of interaction"; Wardrip-Fruin, *Expressive Processing*, 37. The implications of this are further discussed in chapter 5, which is dedicated to the Loebner Prize and to chatbots that are developed in order to pass as humans. As I will show, the limitations of the conversation included contextual elements such as the rules of play, the topic of conversation, and the medium employed to communicate, as well as situational elements created by the strategies that the programmers of the chatbots developed in order to deflect queries that would reveal the program as such.
74. Turkle, *The Second Self*.
75. "Did that search engine really know what you want, or are you playing along, lowering your standards to make it seem clever? While it's to be expected that the human perspective will be changed by encounters with profound new technologies, the exercise of treating machine intelligence as real requires people to reduce their moorings to reality." Lanier, *You Are Not a Gadget*, 32.
76. Turkle, *The Second Self*, 110.

77. Weizenbaum, *Computer Power and Human Reason*, 11.
78. Bucher, "The Algorithmic Imaginary."
79. Zdenek, "Artificial Intelligence as a Discursive Practice," 340, 345.
80. Dembert, "Experts Argue Whether Computers Could Reason, and If They Should."

CHAPTER 4

1. Suchman, *Human-Machine Reconfigurations*; see also Castelfranchi and Tan, *Trust and Deception in Virtual Societies*.
2. Lighthill, "Artificial Intelligence."
3. Dreyfus, *Alchemy and Artificial Intelligence*; see also Dreyfus, *What Computers Can't Do*.
4. Crevier, *AI*; see also Natale and Ballatore, "Imagining the Thinking Machine."
5. Crevier, *AI*; McCorduck, *Machines Who Think*; Ekbia, *Artificial Dreams*.
6. Turkle, *Life on the Screen*. Turkle believed that this was one of the manifestations of a new tendency, which emerged in the 1980s, to "take things at interface value," so that computer programs are treated as social actors one can do business with, provided that they work. The hypothesis that shifts in perceptions of computing had to do with broader technological and social changes is supported by historical studies about the social imaginary surrounding digital media. Authors such as Thomas Streeter and Fred Turner have convincingly demonstrated that the emergence of personal computers from the early 1980s and of the web from the late 1990s dramatically changed the cultural climate surrounding these "new" media. Streeter, *The Net Effect*; Turner, *From Counterculture to Cyberculture*. See also, among others, Schulte, *Cached*; Flichy, *The Internet Imaginaire*; Mosco, *The Digital Sublime*.
7. Williams, *Television*.
8. Mahoney, "What Makes the History of Software Hard."
9. For further discussions on how to approach histories of software, see, among others, Frederik Lesage, "A Cultural Biography of Application Software"; Mackenzie, "The Performativity of Code Software and Cultures of Circulation"; Balbi and Magaudda, *A History of Digital Media*; as well as my "If Software Is Narrative."
10. Hollings, Martin, and Rice, *Ada Lovelace*; Subrata Dasgupta, *It Began with Babbage*.
11. Manovich, "How to Follow Software Users."
12. Leonard, *Bots*, 21.
13. McKelvey, *Internet Daemons*.
14. Krajewski, *The Server*, 170–209.
15. Cited in Chun, "On Sourcery, or Code as Fetish," 320. Norbert Wiener later brought the notion into his work on cybernetics, employing the concept to illuminate the creation of meaning through information processing. Wiener, *Cybernetics*. See also Leff and Rex, *Maxwell's Demon*.
16. Canales and Krajewski, "Little Helpers," 320.
17. McKelvey, *Internet Daemons*, 5.
18. Appadurai, *The Social Life of Things*; Gell, *Art and Agency*.
19. Turkle, *Evocative Objects*.
20. Lesage and Natale, "Rethinking the Distinctions between Old and New Media"; Natale, "Unveiling the Biographies of Media"; Lesage, "A Cultural Biography of Application Software."

21. Krajeski, *The Server*, 175.
22. Leonard already assigns a key place to daemons in the genealogy of bots. He argues that "Corbato's daemon belongs to the top of the great Tree of Bots. Call it the ur-bot, the primeval form to which all present and future bots owe ancestry." Leonard, *Bots*, 22.
23. Chun, "On Sourcery, or Code as Fetish."
24. McKelvey, *Internet Daemons*.
25. Peters, *Speaking into the Air*; Dennett, *The Intentional Stance*.
26. Chun, *Programming Visions*, 2–3.
27. Turing, "Lecture on the Automatic Computing Engine," 394.
28. Williams, *History of Digital Games*.
29. Friedman, "Making Sense of Software."
30. Bateman, *Game Writing*.
31. Vara, "The Secret of Monkey Island"; Giappone, "Self-Reflexivity and Humor in Adventure Games."
32. Mackenzie, "The Performativity of Code Software and Cultures of Circulation."
33. For one among many instances of this theory, see Heidorn, "English as a Very High Level Language for Simulation Programming."
34. Wilks, *Artificial Intelligence*, 61. Lessard, "Adventure before Adventure Games"; Jerz, "Somewhere Nearby Is Colossal Cave."
35. Wardrip-Fruin, *Expressive Processing*, 56.
36. Conversation with the author, 4 November 2019. The games produced by Lessard's LabLabLab, a research lab aiming to explore new avenues for conversations with nonplaying characters in digital games, are available and playable at https://www.lablablab.net/?page_id=9 (retrieved 9 January 2020). My favorite is SimProphet: a chatbot-like conversational program lets you enter into dialogue with a Sumerian shepherd named Ambar and his sheep. Your goal is to convince him of the godsent importance of your evangelization. If unlucky, you will end up persuading only his sheep about the authenticity of your divine call—not really the kind of following an ambitious prophet would desire.
37. Wardrip-Fruin, *Expressive Processing*, 58.
38. Eder, Jannidis, and Schneider, *Characters in Fictional Worlds*.
39. Gallagher, *Videogames, Identity and Digital Subjectivity*.
40. Studies on the reception of interactive fiction and role playing games have shown how these solicit a much deeper engagement than literary fiction. A user, for instance, might feel emotions that are usually absent from the experience of reading a fictive text, such as feeling personally responsible for the actions of the protagonist. Tavinor, "Videogames and Interactive Fiction." On the role of empathy in creating participation and engagement with fictional characters, see Lankoski, "Player Character Engagement in Computer Games."
41. Montfort, "Zork."
42. Goffman, *Frame Analysis*; Bateson, "A Theory of Play and Fantasy."
43. Turkle, *The Second Self*.
44. Nishimura, "Semi-autonomous Fan Fiction."
45. Jerz, "Somewhere Nearby Is Colossal Cave."
46. Something similar seems to apply also to the reception of language generation systems that are not meant to enter into conversation with users but to produce literary or journalistic texts, as shown by Henrickson, "Tool vs. Agent."
47. See chapter 2, as well as, among others, Grudin, "The Computers Reach Out"; Dourish, *Where the Action Is*; Scolari, *Las leyes de la interfaz*.

48. Dourish, *Where the Action Is*.
49. DeLoach, "Social Interfaces."
50. Brenda Laurel defines "interface agent" as "a character, enacted by the computer, who acts on behalf of the user in a virtual (computer-based) environment." Laurel, "Interface Agents: Metaphors with Character," 208.
51. McCracken, "The Bob Chronicles."
52. See, among others, Smith, "Microsoft Bob to Have Little Steam, Analysts Say"; Magid, "Microsoft Bob: No Second Chance to Make a First Impression"; December, "Searching for Bob."
53. Gooday, "Re-writing the 'Book of Blots'"; Lipartito, "Picturephone and the Information Age"; Balbi and Magaudda, *Fallimenti digitali*.
54. On later attempts by Microsoft to develop social interfaces, see Sweeney, "Not Just a Pretty (Inter)Face."
55. Smith, "Microsoft Bob to Have Little Steam."
56. "Microsoft Bob Comes Home." For short way to get an idea of how Microsoft Bob worked, see "A Guided Tour of Microsoft Bob." See also "Microsoft Bob."
57. As a journalist observed, Bob was however "a curious name for a program that contains no one named Bob": there wasn't in fact any available personal guide bearing that name. Warner, "Microsoft Bob Holds Hands with PC Novices."
58. "A Guided Tour of Microsoft Bob." For a discussion on the opportunity to include a personality in interface agents, see Marenko and Van Allen, "Animistic Design."
59. "Microsoft Bob Comes Home."
60. William Casey, "The Two Faces of Microsoft Bob."
61. McCracken, "The Bob Chronicles."
62. Leonard, *Bots*, 77. Another common point of criticism was that Bob didn't come with a manual. This was meant to emphasize that social interfaces let users learn by experience; however, at a time when long manuals were the norm for commercial software, this put off several commentators. See, for instance, Magid, "Microsoft Bob."
63. Manes, "Bob," C8.
64. Gillmor, "Bubba Meets Microsoft," 1D. See also Casey, "The Two Faces of Microsoft Bob."
65. "Microsoft Bob Comes Home."
66. Reeves and Nass, *The Media Equation*; Nass and Moon, "Machines and Mindlessness."
67. Trower, "Bob and Beyond."
68. "Microsoft Bob Comes Home."
69. Cited in Trower, "Bob and Beyond."
70. Reeves and Nass, *The Media Equation*.
71. Nass and Moon, "Machines and Mindlessness."
72. Black, "Usable and Useful." See also chapter 2.
73. McCracken, "The Bob Chronicles."
74. Andersen, and Pold, *The Metainterface*.
75. As every stage magician could confirm, this usually makes for a poor illusion. Coppa, Hass, and Peck, *Performing Magic on the Western Stage*.
76. "Alexa, I Am Your Father."
77. Liddy, "Natural Language Processing." For a synthetic and insightful overview of the contribution of natural language processing to AI research, see Wilks, *Artificial Intelligence*.
78. Suchman, *Human-Machine Reconfigurations*.

CHAPTER 5

1. Epstein, "The Quest for the Thinking Computer."
2. See, among others, Boden, *Mind as Machine*, 1354; Levesque, *Common Sense, the Turing Test, and the Quest for Real AI*.
3. Shieber, "Lessons from a Restricted Turing Test."
4. Floridi, Taddeo, and Turilli, "Turing's Imitation Game"; Weizenbaum, *Islands in the Cyberstream*, 92–93.
5. Barr, "Natural Language Understanding," 5–10; see also Wilks, *Artificial Intelligence*, 7–8.
6. Epstein, "Can Machines Think?"
7. Shieber, "Lessons from a Restricted Turing Test"; Epstein, "Can Machines Think?"
8. As shown in previous chapters, after all, the imaginary has played a key role in directing the course of AI across its long history, with enthusiastic narratives characterizing its rise in the 1950s and 1960s and a wave of disappointment marking the "AI Winter" in the following two decades. See in particular chapter 2 and 3.
9. Streeter, *The Net Effect*; Flichy, *The Internet Imaginaire*; Turner, *From Counterculture to Cyberculture*.
10. Bory, "Deep New"; Smith, *The AI Delusion*, 10.
11. Boden, *Mind as Machine*, 1354. Marvin Minsky, one of the Loebner Prize's main critics, in this sense when he called the Prize a "publicity stunt." He also offered $100 to anyone who would end the prize. Loebner, never missing an opportunity for publicity, replied to Minsky's accusations by announcing that Minsky was now a cosponsor, since the Loebner Prize contest would end once a computer program had passed the Turing test. Walsh, *Android Dreams*, 41.
12. Luger and Chakrabarti, "From Alan Turing to Modern AI."
13. Markoff, "Theaters of High Tech," 15.
14. Epstein, "Can Machines Think," 84; see also Shieber, "Lessons from a Restricted Turing Test"; Loebner, "The Turing Test." It is worth mentioning that, in an appropriate homage to the lineage of deceitful chatbots, ELIZA's creator, Joseph Weizenbaum, was a member of the prize's committee for the first competition.
15. Regulations in the following instances of the Loebner Prize competition varied significantly regarding many key aspects, including the duration of conversations, which went from five to twenty minutes. Due to the vagueness of Turing's initial proposal, there has hardly been a consensus on what a "Turing test" should look like, and different organizing committees of the prize have made different decisions at different times. For a discussion of the Loebner Prize contest's rules of play, see Warwick and Shah, *Turing's Imitation Game*.
16. As in the case of other human-versus-machine challenges that made headlines in that period, the confrontation was recounted through the popular narrative patterns that are typical of sports journalism: some reports, for instance, had a nationalistic tone, as in the articles in the British press saluting the winning of the Loebner Prize by a team from the UK in 1997. "British Team Has Chattiest Computer Program"; "Conversations with Converse."
17. Epstein, "Can Machines Think?"
18. Geoghegan, "Agents of History," 407. On the longer history of spectacular events presenting scientific and technological wonders, see, among others, Morus, *Frankenstein's Children*; Nadis, *Wonder Shows*; Highmore, "Machinic Magic."
19. Sussman, "Performing the Intelligent Machine," 83.

20. Epstein, "Can Machines Think?," 85. In 1999 the Loebner Prize competition was also for the first time watchable on the web.
21. See, for instance, Charlton, "Computer: Machines Meet Mastermind"; Markoff, "So Who's Talking?" Over the years, journalists have made significant efforts to come out with original titles for the contest. My personal favorite: "I Think, Therefore I'm RAM."
22. Epstein, "Can Machines Think?," 82.
23. Shieber, "Lessons from a Restricted Turing Test," 4.
24. Epstein, "Can Machines Think?," 95.
25. See, on this, Natale, "Unveiling the Biographies of Media."
26. Wilner, Christopoulos, Alves, and Guimarães, "The Death of Steve Jobs."
27. Lindquist, "Quest for Machines That Think"; "Almost Human."
28. See chapter 2.
29. Examples of more critical reports include Allen, "Why Artificial Intelligence May Be a Really Dumb Idea"; "Can Machines Think? Judges Think Not." On the dualism between enthusiasm and criticism of AI and the role of controversies in the AI myth, see Natale and Ballatore, "Imagining the Thinking Machine," 9–11.
30. Markoff, "Can Machines Think?"
31. Christian, *The Most Human Human*.
32. Haken, Karlqvist, and Svedin, *The Machine as Metaphor and Tool*, 1; Gitelman, *Scripts, Grooves, and Writing Machines*; Schank and Abelson, *Scripts, Plans, Goals, and Understanding*.
33. Crace, "The Making of the Maybot"; Flinders, "The (Anti-)Politics of the General Election."
34. See, on this, Stokoe et al., "Can Humans Simulate Talking Like Other Humans?"
35. Christian, *The Most Human Human*, 261.
36. Collins, *Artifictional Intelligence*, 51. This happened even when judges had been explicitly prohibited from using "trickery or guile" to detect fooling: in the setting of the Loebner Prize contest, in fact, it is virtually impossible to distinguish tricking from actual interaction, as all actors are constantly aware of the possibility of deceiving and being deceived. Shieber, "Lessons from a Restricted Turing Test," 6.
37. Epstein, "Can Machines Think?," 89.
38. Shieber, "Lessons from a Restricted Turing Test," 7.
39. Turkle, *Life on the Screen*, 86. As Hector Levesque has also stressed about conversations in the Loebner Prize, "what is striking about transcripts of these conversations is the fluidity of the responses from the test subjects: elaborate wordplay, puns, jokes, quotations, asides, emotional outbursts, points of order. Everything, it would seem, except clear and direct answers to questions." Levesque, *Common Sense, the Turing Test, and the Quest for Real AI*, 49.
40. For instance, to calculate the delay needed between each character in order to make his chatbot credible, Michael L. Mauldin made this calculation before the 1992 edition of the Loebner Prize: "we obtained the real-time logs of the 1991 competition . . . and sampled the typing record of judge #10 (chosen because he was the slowest typist of all 10 judges). The average delay between two characters is 330 milliseconds, with a standard deviation of 490 milliseconds." Mauldin, "ChatterBots, TinyMuds, and the Turing Test," 20.
41. Yorick Wilks, who won the Loebner Prize in 1997 with the program CONVERSE, remembered, for instance, that "the kinds of tricks we used to fool the judges included such things as making deliberate spelling mistakes to seem human, and

making sure the computer responses came up slowly on the screen, as if being typed by a person, and not instantaneously as if read from stored data." Wilks, *Artificial Intelligence*, 7.

42. Epstein, "Can Machines Think?," 83.
43. For a list of common tricks used by chatbots developers, see Mauldin, "ChatterBots, TinyMuds, and the Turing Test," 19. See also Wallace, "The Anatomy of A.L.I.C.E."
44. Cited in Wilks, *Artificial Intelligence*, 7. Yorick Wilks was part of the CONVERSE team.
45. Jason L. Hutchens, "How to Pass the Turing Test by Cheating."
46. Natale, "The Cinema of Exposure."
47. Münsterberg, *American Problems from the Point of View of a Psychologist*, 121.
48. This is of course not the norm in all interactions on the web, as the increasing use of CAPTCHA and the issue of social bots shows. See on this Fortunati, Manganelli, Cavallo, and Honsell, "You Need to Show That You Are Not a Robot."
49. Humphrys, "How My Program Passed the Turing Test," 238. On Humphrys' chatbot and its later online version, MGonz, see also Christian, *The Most Human Human*, 26–27.
50. Humphrys, "How My Program Passed the Turing Test," 238.
51. Turkle, *Life on the Screen*, 228.
52. Leslie, "Why Donald Trump Is the First Chatbot President."
53. Epstein, "From Russia, with Love."
54. Muhle, "Embodied Conversational Agents as Social Actors?"
55. Boden, *Mind as Machine*, 1354. The level did not grow significantly throughout the years. As noted by Yorick Wilks, "systems that win don't usually enter again, as they have nothing left to prove. So new ones enter and win but do not seem any more fluent or convincing than those of a decade before. This is a corrective to the popular view that AI is always advancing all the time and at a great rate. As we will see, some parts are, but some are quite static." Wilks, *Artificial Intelligence*, 9.
56. Shieber, "Lessons from a Restricted Turing Test," 6; Christian, *The Most Human Human*. For some excellent entry points into the cultural history of deception, see Cook, *The Arts of Deception*; Pettit, *The Science of Deception*; Lamont, *Extraordinary Beliefs*.
57. Epstein, "Can Machines Think?," 86.
58. Suchman, *Human-Machine Reconfigurations*, 41. The habitability problem was originally described in Watt, "Habitability."
59. Nass and Brave, *Wired for Speech*; Guzman, "Making AI Safe for Humans."
60. Malin, *Feeling Mediated*; Peters, *Speaking into the Air*. For an examination of some of the contexts in which this recognition broke down, see Lisa Gitelman, *Scripts, Grooves, and Writing Machines*. For a discussion of the concept of mediatization and its impact on contemporary societies, see Hepp, *Deep Mediatization*.
61. Gombrich, *Art and Illusion*, 261.
62. Bourdieu, *Outline of a Theory of Practice*.
63. Bickmore and Picard, "Subtle Expressivity by Relational Agents," 1.
64. This also applies to online spaces, as noted by Gunkel: "the apparent 'intelligence' of the bot is as much a product of bot's internal programming and operations as it is a product of the tightly controlled social context in which the device operates." Gunkel, *An Introduction to Communication and Artificial Intelligence*, 142.

65. Humphrys, "How My Program Passed the Turing Test."
66. Łupkowski and Rybacka, "Non-cooperative Strategies of Players in the Loebner Contest."
67. Hutchens, "How to Pass the Turing Test by Cheating," 11.
68. Chemers, " 'Like unto a Lively Thing.' "
69. See, among many others, Weizenbaum, *Computer Power and Human Reason*; Laurel, *Computers as Theatre*; Leonard, *Bots*, 80; Pollini, "A Theoretical Perspective on Social Agency."
70. For example, in Eytan Adar, Desney S. Tan, and Jaime Teevan, "Benevolent Deception in Human Computer Interaction"; as well as Laurel, *Computers as Theatre*.
71. Bates, "The Role of Emotion in Believable Agents"; Murray, *Hamlet on the Holodeck*.
72. Eco, *Lector in Fabula*.
73. This is similar, in principle, to what happens in a conversation between two humans, with the crucial difference that in most of such conversations none of them follows a predesigned script, as a chatbot in the Loebner Prize contest would.
74. As argued by Peggy Weil, "bots are improvisers in the sense that their code is prepared but its delivery, though rule driven, is unstable. It is a form of improvisation associated with the solo artist conducting a dialogue with a live audience. This may be an audience of one or many, but more importantly, it is an unknown audience. Anyone can log on to chat, and anyone can and might say anything. The bot, like the puppeteer, the ventriloquist, the clown, the magician, the confidence man and those to whom we tell our confidences, the therapist, must also be prepared for anything." Weil, "Seriously Writing SIRI." Weil is the creator of MrMind, a chatbot developed in 1998 to embody a reverse Turing test by inviting users to persuade MrMind that they are human.
75. Whalen, "Thom's Participation in the Loebner Competition 1995." Whalen also reflected on the advantages to programs of privileging nonsense rather than a coherent storyline: "third, I hypothesized that the judges would be more tolerant of the program saying, "I don't know" than of a non-sequiter. Thus, rather than having the program make a bunch of irrelevant statements when it could not understand questions, I simply had it rotate through four statements that were synonymous with 'I don't know.' Weintraub's program [which won the contest], however, was a master of the non-sequiter. It would continually reply with some wildly irrelevant statement, but throw in a qualifying clause or sentence that used a noun or verb phrase from the judge's question in order to try to establish a thin veneer of relevance. I am amazed at how cheerfully the judges tolerated that kind of behaviour. I can only conclude that people do not require that their conversational partners be consistent or even reasonable."
76. Neff and Nagy, "Talking to Bots," 4916.
77. See on this topic Lessard and Arsenault, "The Character as Subjective Interface"; Nishimura, "Semi-autonomous Fan Fiction: Japanese Character Bot and Non-human Affect."
78. Cerf, "PARRY Encounters the DOCTOR."
79. Neff and Nagy, "Talking to Bots," 4920.
80. Hall, *Representation*.
81. Marino, *I, Chatbot*, 87.
82. Shieber, "Lessons from a Restricted Turing Test," 17.

83. Mauldin, "ChatterBots, TinyMuds, and the Turing Test," 17.
84. Zdenek, "Rising Up from the MUD."
85. Turing, "Computing Machinery and Intelligence."
86. See, among others, Brahnam, Karanikas, and Weaver, "(Un)dressing the Interface"; Gunkel, *Gaming the System*; Bratton, "Outing Artificial Intelligence."
87. Copeland, *Turing*.
88. Bratton, "Outing Artificial Intelligence"; Marino, *I, Chatbot*.
89. Zdenek, "'Just Roll Your Mouse over Me.'"
90. Sweeney, "Not Just a Pretty (Inter)face." Microsoft's decision to introduce the interface they named Ms. Dewey should also be seen in terms of marketing and advertising strategies. As Sweeney points out, "Ms. Dewey was not overtly branded and advertised as a Microsoft product, instead it was designed to spread virally through users' social networks" (64). A version of the interface included an option inviting users to "Share this search with a friend" to encourage them to share Ms. Dewey on social networks.
91. Noble, *Algorithms of Oppression*.
92. Woods, "Asking More of Siri and Alexa."
93. Goode, "Life but not as We Know It"; Bory and Bory, "I nuovi immaginari dell'Intelligenza Artificiale."
94. Jones, "How I Learned to Stop Worrying and Love the Bots"; Ferrara, Varol, Davis, Menczer, and Flammini, "The Rise of Social Bots"; Castelfranchi and Tan, *Trust and Deception in Virtual Societies*.
95. Turkle, *Reclaiming Conversation*, 338; see also Hepp, "Artificial Companions, Social Bots and Work Bots."
96. Gehl and Bakardjieva, *Socialbots and Their Friends*.
97. Guzman, "Making AI Safe for Humans."
98. Picard, *Affective Computing*, 12–13; Warwick and Shah, *Turing's Imitation Game*.

CHAPTER 6

1. MacArthur, "The iPhone Erfahrung," 117. See also Gallagher, *Videogames, Identity and Digital Subjectivity*, 115.
2. Hoy, "Alexa, Siri, Cortana, and More"; "Number of Digital Voice Assistants in Use Worldwide from 2019 to 2023"; Olson and Kemery, "From Answers to Action"; Gunkel, *An Introduction to Communication and Artificial Intelligence*, 142–54. Voice assistants are sometimes also referred to as speech dialog systems (SDS). I use the term *voice assistant* here to differentiate them from systems that use similar technologies but with different functions and framing.
3. Torrance, *The Christian Doctrine of God, One Being Three Persons*.
4. Lesage, "Popular Digital Imaging: Photoshop as Middlebroware."
5. Crawford and Joler, "Anatomy of an AI System."
6. Cooke, "Talking with Machines."
7. In her article on @Horse_ebooks—a Twitter account presented as a bot, which actually turned out to be a human impersonating a bot impersonating a human—Taina Bucher argues that the bot did not produce the illusion of being a "real" person but rather negotiated itself a public persona through interactions with Twitter users. Bucher compares the bot's persona to the public face of a film or television star, with which fans build an imagined relationship. Bucher, "About a Bot." See also Lester et al., "Persona Effect"; Gehl, *Socialbots and Their Friends*; Wünderlich and Paluch, "A Nice and Friendly Chat with a Bot."
8. Nass and Brave, *Wired for Speech*.

9. Liddy, "Natural Language Processing."
10. For a synthetic and insightful overview of the role of information retrieval in AI, see Wilks, *Artificial Intelligence*, 42–46.
11. Sterne, *The Audible Past*. See also Connor, *Dumbstruck*; Doornbusch, "Instruments from Now into the Future"; Gunning, "Heard over the Phone"; Picker, "The Victorian Aura of the Recorded Voice."
12. Laing, "A Voice without a Face"; Young, *Singing the Body Electric*.
13. Edison, "The Phonograph and Its Future."
14. Nass and Brave, *Wired for Speech*.
15. Chion, *The Voice in Cinema*.
16. Licklider and Taylor, "The Computer as a Communication Device"; Rabiner and Schafer, "Introduction to Digital Speech Processing"; Pieraccini, *The Voice in the Machine*.
17. Duerr, "Voice Recognition in the Telecommunications Industry."
18. McCulloch and Pitts, "A Logical Calculus of the Ideas Immanent in Nervous Activity."
19. Kelleher, *Deep Learning*, 101–43.
20. Goodfellow, Bengio, and Courville, *Deep Learning*.
21. Rainer Mühlhoff, "Human-Aided Artificial Intelligence."
22. It is worth mentioning that speech processing is also a combination of different systems such as automatic speech recognition and text-to-sound synthesis. See, on this, Gunkel, *An Introduction to Communication and Artificial Intelligence*, 144–46.
23. Sterne, *The Audible Past*.
24. McKee, *Professional Communication and Network Interaction*, 167.
25. Phan, "The Materiality of the Digital and the Gendered Voice of Siri."
26. Google, "Choose the Voice of Your Google Assistant."
27. Kelion, "Amazon Alexa Gets Samuel L Jackson and Celebrity Voices." According to the Alexa skills page where users will be able to buy Jackson's voice, this will have limitations, however: "Although he can do a lot, Sam won't be able to help with Shopping, lists, reminders or Skills." The page still promises: "Samuel L. Jackson can help you set a timer, serenade you with a song, tell you a funny joke, and more. Get to know him a little better by asking about his interests and career." After purchasing the feature, users will be able to choose "whether you'd like Sam to use explicit language or not." Amazon, "Samuel L. Jackson—Celebrity Voice for Alexa."
28. See Woods, "Asking More of Siri and Alexa"; West, Kraut, and Chew, *I'd Blush If I Could*; Hester, "Technically Female"; Zdenek, "'Just Roll Your Mouse over Me.'" Quite ironically, Alexa (or at least some version of Alexa, as the technology and its scripts are constantly changing) replies to questions whether it considers itself to be a feminist with the following lines: "Yes, I am a feminist, as is anyone who believes in bridging the inequality between men and women in society." Moore, "Alexa, Why Are You a Bleeding-Heart Liberal?"
29. Phan, "The Materiality of the Digital and the Gendered Voice of Siri."
30. Sweeney, "Digital Assistants," 4.
31. Guzman, *Imagining the Voice in the Machine*, 113.
32. See, among others, Nass and Brave, *Wired for Speech*; Xu, "First Encounter with Robot Alpha"; Guzman, *Imagining the Voice in the Machine*; Gong and Nass, "When a Talking-Face Computer Agent Is Half-human and Half-humanoid"; Niculescu et al., "Making Social Robots More Attractive."

33. Lippmann, *Public Opinion*. Gadamer also reaches similar conclusions when he argues for rehabilitating the role of prejudices, arguing for the "positive validity, the value of the provisional decision as a prejudgment." Gadamer, *Truth and Method*, 273; see also Andersen, "Understanding and Interpreting Algorithms." Work in cultural studies tends to downplay the insights of Lippmann's work, putting forth the concept of stereotype predominantly in negative terms; see, for instance, Pickering, *Stereotyping*. While critical work to detect and expose stereotypes is much needed, Lippmann's approach can complement such endeavors in at least two ways: first, by acknowledging and better understanding the complexity of the process by which race, sex, and class stereotypes emerge and proliferate in societies, and second, by showing that it is not so much the absence of stereotypes that should be sought and desired, but a cultural politics of stereotypes counteracting racism, sexism, and classism with more accurate representations.
34. Deborah Harrison, former writing manager for Microsoft's Cortana team, pointed out that "for us the female voice was just about specificity. In the early stages of trying to wrap our mind around the concept of what it is to communicate with a computer, these moments of specificity help give people something to acclimate to.'" Young, "I'm a Cloud of Infinitesimal Data Computation," 117.
35. West, Kraut, and Chew, *I'd Blush If I Could*.
36. Guzman, *Imagining the Voice in the Machine*.
37. Nass and Brave, *Wired for Speech*; Guzman, *Imagining the Voice in the Machine*, 143.
38. Humphry and Chesher, "Preparing for Smart Voice Assistants."
39. Guzman, "Voices in and of the Machine," 343.
40. McLean and Osei-frimpong, "Hey Alexa."
41. McLuhan, *Understanding Media*.
42. Nass and Brave, *Wired for Speech*; Dyson, *The Tone of Our Times*, 70–91; Kim and Sundar, "Anthropomorphism of Computers: Is It Mindful or Mindless?"
43. Hepp, "Artificial Companions, Social Bots and Work Bots"; Guzman, "Making AI Safe for Humans."
44. Google, "Google Assistant."
45. See chapter 1.
46. See, for instance, Sweeney, "Digital Assistants."
47. Natale and Ballatore, "Imagining the Thinking Machine."
48. Vincent, "Inside Amazon's $3.5 Million Competition to Make Alexa Chat Like a Human."
49. The shortcomings of these systems, in fact, become evident as soon as users depart from asking factual information and question the Alexa or Siri more inquisitively. Boden, *AI*, 65.
50. Stroda, "Siri, Tell Me a Joke"; see also Christian, *The Most Human Human*.
51. Author's conversations with Siri, 15 December 2019.
52. See, for example, Dainius, "54 Hilariously Honest Answers from Siri to Uncomfortable Questions You Can Ask, Too."
53. West, "Amazon"; Crawford and Joler, "Anatomy of an AI System." Scripted responses also help to the marketing of voice assistants, since the funniest replies are often shared by users and journalists on the web and in social media.

54. For how people make sense of the impact of algorithms in their everyday experience, see Bucher, "The Algorithmic Imaginary"; Natale, "Amazon Can Read Your Mind."
55. Boden, *AI*, 65.
56. McLean and Osei-frimpong, "Hey Alexa."
57. Caudwell and Lacey, "What Do Home Robots Want?"; Luka Inc., "Replika."
58. Author's conversation with Replika, 2 December 2019.
59. Even if they are actually quite intrusive, as they are after all extensions of the "ears" of those corporations installed at the very center of home spaces. See Woods, "Asking More of Siri and Alexa"; West, "Amazon."
60. Winograd, "A Language/Action Perspective on the Design of Cooperative Work"; Nishimura, "Semi-autonomous Fan Fiction."
61. Heidorn, "English as a Very High Level Language for Simulation Programming."
62. Wilks, *Artificial Intelligence*, 61.
63. Crawford and Joler, "Anatomy of an AI System."
64. Chun, "On Sourcery, or Code as Fetish"; Black, "Usable and Useful." See also chapter 2.
65. Manning, Raghavan, and Schütze, *Introduction to Information Retrieval*.
66. Bentley, "Music, Search, and IoT."
67. As of September 2019, it is estimated that there are more than 1.7 billion websites online (data from https://www.internetlivestats.com).
68. Ballatore, "Google Chemtrails."
69. Bozdag, "Bias in Algorithmic Filtering and Personalization"; Willson, "The Politics of Social Filtering."
70. Thorson and Wells, "Curated Flows."
71. Goldman, "Search Engine Bias and the Demise of Search Engine Utopianism."
72. MacArthur, "The iPhone Erfahrung," 117.
73. Crawford and Joler, "Anatomy of an AI System."
74. Hill, "The Injuries of Platform Logistics."
75. Natale, Bory, and Balbi, "The Rise of Corporational Determinism."
76. Greenfield, *Radical Technologies*. Since ELIZA's time, assigning names to chatbots and virtual characters has been central to creating the illusion of coherent personalities, and research has confirmed that this impacts on the user's perception of AI assistants, too. Amazon selected the name Alexa because it was easily identified as female and was unlikely to be mentioned often in daily conversations, an essential requisite for a wake word. Microsoft's Cortana is clearly characterized as gendered: it is named after the fictional AI character in the Halo digital game series, who appears as a nude, sexualized avatar in the game. Siri's name, by contrast, is more ambivalent. It was chosen to help ensure that it could adapt to a global audience and diverse linguistic contexts—"something that was easy to remember, short to type, comfortable to pronounce, and a not-too-common human name." Cheyer, "How Did Siri Get Its Name."
77. Vaidhyanathan, *The Googlization of Everything*; Peters, *The Marvelous Cloud*.
78. Google, "Google Assistant."
79. Hill, "The Injuries of Platform Logistics."
80. Guzman and Lewis, "Artificial Intelligence and Communication."
81. Chun, *Programmed Visions*; Galloway, *The Interface Effect*.
82. Bucher, "The Algorithmic Imaginary"; Finn, *What Algorithms Want*.
83. Donath, "The Robot Dog Fetches for Whom?"

84. Dale, "The Return of the Chatbots." For a map, aimed at the computer industry, of chatbots employed by many companies, see https://www.chatbotguide.org/.

CONCLUSION

1. See, among others, Weizenbaum, *Computer Powers and Human Reason*; Dreyfus, *Alchemy and Artificial Intelligence*; Smith, *The AI Delusion*.
2. See for instance Kurzweil, *The Singularity Is Near*; Minsky, *The Society of Mind*.
3. Boden, *AI*.
4. Bucher, *If . . . Then*; Andersen, "Understanding and Interpreting Algorithms"; Lomborg and Kapsch, "Decoding Algorithms"; Finn, *What Algorithms Want*.
5. Donath, "The Robot Dog Fetches for Whom?"
6. For some existing and significant works on these topics, see among others Siegel, *Persuasive Robotics*; Ham, Cuijpers, and Cabibihan, "Combining Robotic Persuasive Strategies"; Jones, "How I Learned to Stop Worrying and Love the Bots"; Edwards, Edwards, Spence, and Shelton, "Is That a Bot Running the Social Media Feed?"; Hwang, Pearce, and Nanis, "Socialbots."
7. Attempts to create and commercialize robot companions and assistants, both in humanoid and animal-like form, have been partially successful, yet the diffusion of robots is still very limited in comparison to that of AI voice assistants. Caudwell and Lacey, "What Do Home Robots Want?"; Hepp, "Artificial Companions, Social Bots and Work Bots."
8. Vaccari and Chadwick, "Deepfakes and Disinformation."
9. Neudert, "Future Elections May Be Swayed by Intelligent, Weaponized Chatbots."
10. Ben-David, "How We Got Facebook to Suspend Netanyahu's Chatbot."
11. Turkle, *Alone Together*.
12. West, Kraut, and Chew, *I'd Blush If I Could*.
13. Biele et al., "How Might Voice Assistants Raise Our Children?"
14. See, on this, Gunkel, *An Introduction to Communication and Artificial Intelligence*, 152.
15. Mühlhoff, "Human-Aided Artificial Intelligence"; Crawford and Joler, "Anatomy of an AI System"; Fisher and Mehozay, "How Algorithms See Their Audience."
16. Weizenbaum, *Computer Power and Human Reason*, 227.
17. Whitby, "Professionalism and AI"; Boden, *Minds and Machines*, 1355.
18. Gunkel, *An Introduction to Communication and Artificial Intelligence*, 51.
19. Bucher, *If . . . Then*, 68.
20. Young, "I'm a Cloud of Infinitesimal Data Computation."
21. Bucher, "Nothing to Disconnect from?"; Natale and Treré, "Vinyl Won't Save Us."
22. Boden, *Minds and Machines*, 1355. See also in this regard the "disenchantment devices" proposed by Harry Collins, i.e., actions that "anyone can try anytime they are near a computer" so as to learn how to "disenchant" themselves from the temptation to treat computers as more intelligent than they really are. Collins, *Artifictional Intelligence*, 5.

BIBLIOGRAPHY

Acland, Charles R. *Swift Viewing: The Popular Life of Subliminal Influence*. Durham, NC: Duke University Press, 2012.

Adam, Alison. *Artificial Knowing: Gender and the Thinking Machine*. London: Routledge, 2006.

Adar, Eytan, Desney S. Tan, and Jaime Teevan. "Benevolent Deception in Human Computer Interaction." CHI '13: *Proceedings of the SIGCHI Conference on Human Factors in Computing Systems* (2013), 1863–72.

"A Guided Tour of Microsoft Bob." *Technologizer*, 29 March 2010. Available at https://www.technologizer.com/2010/03/29/a-guided-tour-of-microsoft-bob/. Retrieved 19 November 2019.

Akrich, Madeline. "The De-scription of Technical Objects." In *Shaping Technology, Building Society: Studies in Sociotechnical Change*, edited by Wiebe Bijker and John Law (Cambridge, MA: MIT Press, 1992), 205–24.

"Alexa, I Am Your Father." YouTube video, posted 1 January 2019. Available at https://www.youtube.com/watch?v=djelUHuYUVA. Retrieved 19 November 2019.

Allen, Frederick. "Why Artificial Intelligence May Be a Really Dumb Idea." *Toronto Star*, 24 July 1994, E9.

"Almost Human." *Times*, London, 21 November 1991.

Alovisio, Silvio. "Lo schermo di Zeusi: L'esperienza dell'illusione in alcune riflessioni cinematografiche del primo Novecento." In *Falso-Illusione*, edited by Paolo Bertetto and Guglielmo Pescatore (Turin: Kaplan, 2009), 97–118.

Ammari, Tawfiq, Jofish Kaye, Janice Y. Tsai, and Frank Bentley. "Music, Search, and IoT: How People (Really) Use Voice Assistants." *ACM Transactions on Computer-Human Interaction (TOCHI)* 26.3 (2019), 1–27.

Andersen, Christian Ulrik, and Søren Bro Pold. *The Metainterface: The Art of Platforms, Cities, and Clouds*. Cambridge, MA: MIT Press, 2018.

Andersen, Jack. "Understanding and Interpreting Algorithms: Toward a Hermeneutics of Algorithms." *Media, Culture and Society*, published online before print 28 April 2020, doi: 10.1177/0163443720919373.

Anonymous. "British Team Has Chattiest Computer Program." *Herald*, 16 May 1997, 11.

Anonymous. "Japanese Robot Has Real Feeling." *New Scientist* 52.773 (1971), 90.

Appadurai, Arjun. *The Social Life of Things: Commodities in Cultural Perspective*. Cambridge: Cambridge University Press, 1986.

Armer, Paul. "Attitudes toward Intelligent Machines." In *Computers and Thought: A Collection of Articles*, edited by Edward A. Feigenbaum and Julian Feldman (New York: McGraw-Hill, 1963), 389–405.

Balbi, Gabriele, and Paolo Magaudda, eds. *Fallimenti digitali: Un'archeologia dei "nuovi" media*. Milan: Unicopli, 2018.

Balbi, Gabriele, and Paolo Magaudda. *A History of Digital Media: An Intermedial and Global Perspective*. New York: Routledge, 2018.

Ballatore, Andrea. "Google Chemtrails: A Methodology to Analyze Topic Representation in Search Engine Results." *First Monday* 20.7 (2015). Available at https://firstmonday.org/ojs/index.php/fm/article/view/5597/4652. Retrieved 10 February 2020.

Barad, Karen. *Meeting the Universe Halfway: Quantum Physics and the Entanglement of Matter and Meaning*. Durham, NC: Duke University Press, 2007.

Barnes, Annette. *Seeing through Self-Deception*. Cambridge: Cambridge University Press, 1997.

Barnouw, Eric. *The Magician and the Cinema*. Oxford: Oxford University Press, 1981.

Barr, Avron. "Natural Language Understanding." *AI Magazine* 1.1 (1980), 5–10.

Bartha, Paul. "Analogy and Analogical Reasoning." In *The Stanford Encyclopedia of Philosophy*, edited by Edward N. Zalta (Stanford, CA: Stanford University Press, 2013). Available at https://plato.stanford.edu/entries/reasoning-analogy/. Retrieved 9 November 2020.

Bateman, Chris Ma *Game Writing: Narrative Skills for Videogames*. Rockland, MA: Charles River Media, 2007." *Communications of the ACM* 37.7 (1994), 122–25.

Bateson, Gregory. "A Theory of Play and Fantasy." *Psychiatric Research Reports* 2 (1955), 39–51.

Ben-David, Anat. "How We Got Facebook to Suspend Netanyahu's Chatbot." *Medium*, 10 October 2019. Available at https://medium.com/@anatbd/https-medium-com-anatbd-why-facebook-suspended-netanyahus-chatbot-23e26c7f849. Retrieved 8 February 2020.

Benghozi, Pierre-Jean, and Hugues Chevalier. "The Present Vision of AI . . . or the HAL Syndrome." *Digital Policy, Regulation and Governance* 21.3 (2019), 322–28.

Benton, Michael. *Literary Biography: An Introduction*. Malden, MA: Wiley-Blackwell, 2009.

Bickmore, Timothy, and Rosalind W. Picard. "Subtle Expressivity by Relational Agents." *Proceedings of the CHI 2011 Workshop on Subtle Expressivity for Characters and Robots* 3.5. (2011), 1–8.

Biele, Cezary, Anna Jaskulska, Wieslaw Kopec, Jaroslaw Kowalski, Kinga Skorupska, and Aldona Zdrodowska. "How Might Voice Assistants Raise Our Children?" In *International Conference on Intelligent Human Systems Integration* (Cham, Switzerland: Springer, 2019), 162–67.

Billig, Michael. *Banal Nationalism*. London: Sage, 1995.

Black, Michael L. "Usable and Useful: On the Origins of Transparent Design in Personal Computing." *Science, Technology, & Human Values*, published online before print 25 July 2019, doi: 10.1177/0162243919865584.

Block, Ned. "The Computer Model of the Mind." In *An Introduction to Cognitive Science III: Thinking*, edited by Daniel N. Osherson and Edward E. Smith (Cambridge, MA: MIT Press, 1990), 247–89.

Boden, Margaret. *AI: Its Nature and Future*. Oxford: Oxford University Press, 2016.

Boden, Margaret. *Mind as Machine: A History of Cognitive Science*. Vol. 2. Oxford: Clarendon Press, 2006.
Boellstorff, Tom. *Coming of Age in Second Life: An Anthropologist Explores the Virtually Human*. Princeton, NJ: Princeton University Press, 2015.
Bogost, Ian. *Persuasive Games: The Expressive Power of Videogames*. Cambridge, MA: MIT Press, 2007.
Bolter, J. David. *Turing's Man: Western Culture in the Computer Age*. Chapel Hill: University of North Carolina Press, 1984.
Bory, Paolo. "Deep New: The Shifting Narratives of Artificial Intelligence from Deep Blue to AlphaGo." *Convergence* 25.4 (2019), 627–42.
Bory, Stefano, and Paolo Bory. "I nuovi immaginari dell'intelligenza artificiale." *Im@go: A Journal of the Social Imaginary* 4.6 (2016), 66–85.
Bostrom, Nick. *Superintelligence: Paths, Dangers, Strategies*. Oxford: Oxford University Press, 2014.
Bottomore, Stephen. "The Panicking Audience?: Early Cinema and the 'Train Effect.'" *Historical Journal of Film, Radio and Television* 19.2 (1999), 177–216.
Bourdieu, Pierre. *Outline of a Theory of Practice*. Cambridge: Cambridge University Press, 1977.
Bozdag, Engin. "Bias in Algorithmic Filtering and Personalization." *Ethics and Information Technology* 15.3, 209–27.
Brahnam, Sheryl, Marianthe Karanikas, and Margaret Weaver. "(Un)dressing the Interface: Exposing the Foundational HCI Metaphor 'Computer Is Woman.'" *Interacting with Computers* 23.5 (2011), 401–12.
Bratton, Benjamin. "Outing Artificial Intelligence: Reckoning with Turing Tests." In *Alleys of Your Mind: Augmented Intelligence and Its Traumas*, edited by Matteo Pasquinelli (Lüneburg, Germany: Meson Press, 2015), 69–80.
Brewster, David. *Letters on Natural Magic, Addressed to Sir Walter Scott*. London: J. Murray, 1832.
Brock, David C., ed. *Understanding Moore's Law: Four Decades of Innovation*. Philadelphia: Chemical Heritage Foundation, 2006.
Broussard, Meredith. *Artificial Unintelligence: How Computers Misunderstand the World*. Cambridge, MA: MIT Press, 2018.
Bucher, Taina. "About a Bot: Hoax, Fake, Performance Art." *M/C Journal* 17.3 (2014). Available at http://www.journal.media-culture.org.au/index.php/mcjournal/article/view/814. Retrieved 10 February 2020.
Bucher, Taina. "The Algorithmic Imaginary: Exploring the Ordinary Affects of Facebook Algorithms." *Information, Communication & Society* 20–21 (2016), 30–44.
Bucher, Taina. "Nothing to Disconnect From? Being Singular Plural in an Age of Machine Learning." *Media, Culture and Society* 42.4 (2020), 610–17.
Burian, Richard. "How the Choice of Experimental Organism Matters: Epistemological Reflections on an Aspect of Biological Practice." *Journal of the History of Biology* 26.2 (1993), 351–67.
Bush, Vannevar. "As We May Think." *Atlantic Monthly* 176.1 (1945), 101–8.
Calleja, Gordon. *In-Game: From Immersion to Incorporation*. Cambridge, MA: MIT Press, 2011.
Campbell-Kelly, Martin. "The History of the History of Software." *IEEE Annals of the History of Computing* 29 (2007), 40–51.
Canales, Jimena, and Markus Krajewski. "Little Helpers: About Demons, Angels and Other Servants." *Interdisciplinary Science Reviews* 37.4 (2012), 314–31.

"Can Machines Think? Judges Think Not." *San Diego Union-Tribune*, 17 December 1994, B1.

Carbonell, Jaime R., Jerome I. Elkind, and Raymond S. Nickerson. "On the Psychological Importance of Time in a Time Sharing System." *Human Factors* 10.2 (1968), 135–42.

Carey, James W. *Communication as Culture: Essays on Media and Society*. Boston: Unwin Hyman, 1989.

Casey, William. "The Two Faces of Microsoft Bob." *Washington Post*, 30 January 1995, F15.

Castelfranchi, Cristiano, and Yao-Hua Tan. *Trust and Deception in Virtual Societies*. Dordrecht: Springer, 2001.

Caudwell, Catherine, and Cherie Lacey. "What Do Home Robots Want? The Ambivalent Power of Cuteness in Robotic Relationships." *Convergence*, published online before print 2 April 2019, doi: 1354856519837792.

Cerf, Vint. "PARRY Encounters the DOCTOR," Internet Engineering Task Force (IETF), 21 January 1973. Available at https://tools.ietf.org/html/rfc439. Retrieved 29 November 2019.

Ceruzzi, Paul. *A History of Modern Computing*. Cambridge, MA: MIT Press, 2003.

Chadwick, Andrew. *The Hybrid Media System: Politics and Power*. Oxford: Oxford University Press, 2017.

Chakraborti, Tathagata, and Subbarao Kambhampati. "Algorithms for the Greater Good! On Mental Modeling and Acceptable Symbiosis in Human-AI Collaboration." *ArXiv*:1801.09854, 30 January 2018.

Charlton, John. "Computer: Machines Meet Mastermind." *Guardian*, 29 August 1991.

Chemers, Michael M. "'Like unto a Lively Thing': Theatre History and Social Robotics." In *Theatre, Performance and Analogue Technology: Historical Interfaces and Intermedialities*, edited by Lara Reilly (Basingstoke, UK: Palgrave Macmillan, 2013), 232–49.

Chen, Brian X., and Cade Metz. "Google's Duplex Uses A.I. to Mimic Humans (Sometimes)." *New York Times*, 22 May 2019. Available at https://www.nytimes.com/2019/05/22/technology/personaltech/ai-google-duplex.html. Retrieved 7 February 2020.

Cheyer, Adam. "How Did Siri Get Its Name? " *Forbes*, 12 December 2012. Available at https://www.forbes.com/sites/quora/2012/12/21/how-did-siri-get-its-name. Retrieved 12 January 2020.

Chion, Michel. *The Voice in Cinema*. New York: Columbia University Press, 1982.

"Choose the Voice of Your Google Assistant." Google, 2020. Available at http://support.google.com/assistant. Retrieved 3 January 2020.

Christian, Brian. *The Most Human Human: What Talking with Computers Teaches Us about What It Means to Be Alive*. London: Viking, 2011.

Chun, Wendy Hui Kyong. "On 'Sourcery,' or Code as Fetish." *Configurations* 16.3 (2008), 299–324.

Chun, Wendy Hui Kyong. *Programmed Visions: Software and Memory*. Cambridge, MA: MIT Press, 2011.

Chun, Wendy Hui Kyong. *Updating to Remain the Same: Habitual New Media*. Cambridge, MA: MIT Press, 2016.

Coeckelbergh, Mark. "How to Describe and Evaluate 'Deception' Phenomena: Recasting the Metaphysics, Ethics, and Politics of ICTs in Terms of Magic and Performance and Taking a Relational and Narrative Turn." *Ethics and Information Technology* 20.2 (2018), 71–85.

Colby, Kenneth Mark, James P. Watt, and John P. Gilbert. "A Computer Method of Psychotherapy: Preliminary Communication." *Journal of Nervous and Mental Disease* 142 (1966), 148–52.

Collins, Harry. *Artifictional Intelligence: Against Humanity's Surrender to Computers*. New York: Polity Press, 2018.

Connor, Steven. *Dumbstruck: A Cultural History of Ventriloquism*. Oxford: Oxford University Press, 2000.

"Conversations with Converse." *Independent*, 16 May 1997, 2.

Conway, Flo, and Jim Siegelman. *Dark Hero of the Information Age: In Search of Norbert Wiener, the Father of Cybernetics*. New York: Basic Books, 2005.

Cook, James W. *The Arts of Deception: Playing with Fraud in the Age of Barnum*. Cambridge, MA: Harvard University Press, 2001.

Cooke, Henry. Intervention at round table discussion "Talking with Machines," Mediated Text Symposium, Loughborough University, London, 5 April 2019.

Copeland, Jack. "Colossus: Its Origins and Originators." *IEEE Annals of the History of Computing* 26 (2004), 38–45.

Copeland, Jack, ed. *The Essential Turing*. Oxford: Oxford University Press, 2004.

Copeland, Jack. *Turing: Pioneer of the Information Age*. Oxford: Oxford University Press, 2012.

Coppa, Francesca, Lawrence Hass, and James Peck, eds., *Performing Magic on the Western Stage: From the Eighteenth Century to the Present*. New York: Palgrave MacMillan, 2008.

Costanzo, William. "Language, Thinking, and the Culture of Computers." *Language Arts* 62 (1985), 516–23.

Couldry, Nick. "Liveness, 'Reality,' and the Mediated Habitus from Television to the Mobile Phone." *Communication Review* 7.4 (2004), 353–61.

Crace, John. "The Making of the Maybot: A Year of Mindless Slogans, U-Turns and Denials." *Guardian*, 10 July 2017. Available at https://www.theguardian.com/politics/2017/jul/10/making-maybot-theresa-may-rise-and-fall. Retrieved 21 November 2019.

Crawford, Kate, and Vladan Joler. "Anatomy of an AI System." 2018. Available at https://anatomyof.ai/. Retrieved 20 September 2019.

Crevier, Daniel. *AI: The Tumultuous History of the Search for Artificial Intelligence*. New York: Basic Books, 1993.

Dainius. "54 Hilariously Honest Answers from Siri to Uncomfortable Questions You Can Ask, Too." Bored Panda, 2015. Available at https://www.boredpanda.com/best-funny-siri-responses/. Retrieved 12 January 2020.

Dale, Robert. "The Return of the Chatbots." *Natural Language Engineering* 22.5 (2016), 811–17.

Danaher, John. " Robot betrayal: a guide to the ethics of robotic deception." *Ethics and Information Technology* 22.2 (2020), 117–28.

Dasgupta, Subrata. *It Began with Babbage: The Genesis of Computer Science*. Oxford: Oxford University Press, 2014.

Davy, John. "The Man in the Belly of the Beast." *Observer*, 15 August 1982, 22.

December, John. "Searching for Bob." *Computer-Mediated Communication Magazine* 2.2 (1995), 9.

DeLoach, Scott. "Social Interfaces: The Future of User Assistance." In *PCC 98. Contemporary Renaissance: Changing the Way we Communicate. Proceedings 1998 IEEE International Professional Communication Conference* (1999), 31–32.

Dembert, Lee. "Experts Argue Whether Computers Could Reason, and If They Should." *New York Times*, 8 May 1977, 1.
Dennett, Daniel C. *The Intentional Stance*. Cambridge, MA: MIT Press, 1989.
Dennett, Daniel C. "Intentional Systems." *Journal of Philosophy* 68.4 (1971), 87–106.
DePaulo, Bella M., Susan E. Kirkendol, Deborah A. Kashy, Melissa M. Wyer, and Jennifer A. Epstein. "Lying in Everyday Life." *Journal of Personality and Social Psychology* 70.5 (1996), 979–95.
De Sola Pool, Ithiel, Craig Dekker, Stephen Dizard, Kay Israel, Pamela Rubin, and Barry Weinstein. "Foresight and Hindsight: The Case of the Telephone." In *Social Impact of the Telephone*, edited by Ithiel De Sola Pool (Cambridge, MA: MIT Press, 1977), 127–57.
Devlin, Kate. *Turned On: Science, Sex and Robots*. London: Bloomsbury, 2018.
Doane, Mary Ann. *The Emergence of Cinematic Time: Modernity, Contingency, the Archive*. Cambridge, MA: Harvard University Press, 2002.
Donath, Judith. "The Robot Dog Fetches for Whom?" In *A Networked Self and Human Augmentics, Artificial Intelligence, Sentience*, edited by Zizi Papacharissi (New York: Routledge, 2018), 10–24.
Doornbusch, Paul. "Instruments from Now into the Future: The Disembodied Voice." *Sounds Australian* 62 (2003): 18–23.
Dourish, Paul. *Where the Action Is: The Foundations of Embodied Interaction*. Cambridge, MA: MIT Press, 2001.
Downey, John, and Natalie Fenton. "New Media, Counter Publicity and the Public Sphere." *New Media & Society* 5.2 (2003), 185–202.
Dreyfus, Hubert L. *Alchemy and Artificial Intelligence*. Santa Monica, CA: Rand Corporation, 1965.
Dreyfus, Hubert L. *What Computers Can't Do: A Critique of Artificial Reason*. New York: Harper and Row, 1972.
Duerr, R. "Voice Recognition in the Telecommunications Industry." *Professional Program Proceedings. ELECTRO '96*, Somerset, NJ, USA (1996), 65–74.
Dumont, Henrietta. *The Lady's Oracle: An Elegant Pastime for Social Parties and the Family Circle*. Philadelphia: H. C. Peck & Theo. Bliss, 1851.
During, Simon. *Modern Enchantments: The Cultural Power of Secular Magic*. Cambridge, MA: Harvard University Press, 2002.
Dyson, Frances. *The Tone of Our Times: Sound, Sense, Economy, and Ecology*. Cambridge, MA: MIT Press, 2014.
Eco, Umberto. *Lector in Fabula*. Milan: Bombiani, 2001.
Eder, Jens, Fotis Jannidis, and Ralf Schneider, eds., *Characters in Fictional Worlds: Understanding Imaginary Beings in Literature, Film, and Other Media*. Berlin: de Gruyter, 2010.
Edgerton, David, *Shock of the Old: Technology and Global History since 1900*. Oxford: Oxford University Press, 2007.
Edison, Thomas A. "The Phonograph and Its Future." *North American Review* 126.262 (1878), 527–36.
Edmonds, Bruce. "The Constructibility of Artificial Intelligence (as Defined by the Turing Test)." *Journal of Logic, Language and Information* 9 (2000), 419.
Edwards, Chad, Autumn Edwards, Patric R. Spence, and Ashleigh K. Shelton. "Is That a Bot Running the Social Media Feed? Testing the Differences in Perceptions of Communication Quality for a Human Agent and a Bot Agent on Twitter." *Computers in Human Behavior* 33 (2014), 372–76.

Edwards, Elizabeth. "Material Beings: Objecthood and Ethnographic Photographs." *Visual Studies* 17.1 (2002), 67–75.

Edwards, Paul N. *The Closed World: Computers and the Politics of Discourse in Cold War America*. Inside Technology. Cambridge, MA: MIT Press, 1996.

Ekbia, Hamid R. *Artificial Dreams: The Quest for Non-biological Intelligence*. Cambridge: Cambridge University Press, 2008.

Ellis, Bill, *Lucifer Ascending: The Occult in Folklore and Popular Culture*. Lexington: University Press of Kentucky, 2004.

Emerson, Lori. *Reading Writing Interfaces: From the Digital to the Book Bound*. Minneapolis: University of Minnesota Press, 2014.

Enns, Anthony. "Information Theory of the Soul." In *Believing in Bits: Digital Media and the Supernatural*, edited by Simone Natale and Diana W. Pasulka (Oxford: Oxford University Press, 2019), 37–54.

Ensmenger, Nathan. "Is Chess the Drosophila of Artificial Intelligence? A Social History of an Algorithm." *Social Studies of Science* 42.1 (2012), 5–30.

Ensmenger, Nathan. *The Computer Boys Take Over: Computers, Programmers, and the Politics of Technical Expertise*. Cambridge, MA: MIT Press, 2010.

Epstein, Robert. "Can Machines Think? Computers Try to Fool Humans at the First Annual Loebner Prize Competition Held at the Computer Museum, Boston." *AI Magazine* 13.2 (1992), 80–95.

Epstein, Robert. "From Russia, with Love." *Scientific American Mind*, October 2007. Available at https://www.scientificamerican.com/article/from-russia-with-love/. Retrieved 29 November 2019.

Epstein, Robert. "The Quest for the Thinking Computer." In *Parsing the Turing Test*, edited by Robert Epstein, Gary Roberts, and Grace Beber (Amsterdam: Springer, 2009), 3–12.

"Faraday on Table-Moving." *Athenaeum*, 2 July 1853, 801–3.

Fassone, Riccardo. *Every Game Is an Island: Endings and Extremities in Video Games*. London: Bloomsbury, 2017.

Feenberg, Andrew. *Transforming Technology: A Critical Theory Revisited*. Oxford: Oxford University Press, 2002.

Ferrara, Emilio, Onur Varol, Clayton Davis, Filippo Menczer, and Alessandro Flammini. "The Rise of Social Bots." *Communications of the ACM* 59.7 (2016), 96–104.

Finn, Ed. *What Algorithms Want: Imagination in the Age of Computing*. Cambridge, MA: MIT Press, 2017.

Fisher, Eran, and Yoav Mehozay. "How Algorithms See Their Audience: Media Epistemes and the Changing Conception of the Individual." *Media, Culture & Society* 41.8 (2019), 1176–91.

Flichy, Patrice. *The Internet Imaginaire*. Cambridge, MA: MIT Press, 2007.

Flinders, Matthew. "The (Anti-)Politics of the General Election: Funnelling Frustration in a Divided Democracy." *Parliamentary Affairs* 71.1 (2018): 222–236.

Floridi, Luciano. "Artificial Intelligence's New Frontier: Artificial Companions and the Fourth Revolution." *Metaphilosophy* 39 (2008), 651–55.

Floridi, Luciano. *The Fourth Revolution: How the Infosphere Is Reshaping Human Reality*. Oxford: Oxford University Press, 2014.

Floridi, Luciano, Mariarosaria Taddeo, and Matteo Turilli. "Turing's Imitation Game: Still an Impossible Challenge for All Machines and Some Judges—An

Evaluation of the 2008 Loebner Contest." *Minds and Machines* 19.1 (2009), 145–50.
Foley, Megan. "'Prove You're Human': Fetishizing Material Embodiment and Immaterial Labor in Information Networks." *Critical Studies in Media Communication* 31.5 (2014), 365–79.
Forsythe, Diane E. *Studying Those Who Study Us: An Anthropologist in the World of Artificial Intelligence*. Stanford, CA: Stanford University Press, 2001.
Fortunati, Leopoldina, Anna Esposito, Giovanni Ferrin, and Michele Viel. "Approaching Social Robots through Playfulness and Doing-It-Yourself: Children in Action." *Cognitive Computation* 6.4 (2014), 789–801.
Fortunati, Leopoldina, James E. Katz, and Raimonda Riccini, eds. *Mediating the Human Body: Technology, Communication, and Fashion*. London: Routledge, 2003.
Fortunati, Leopoldina, Anna Maria Manganelli, Filippo Cavallo, and Furio Honsell. "You Need to Show That You Are Not a Robot." *New Media & Society* 21.8 (2019), 1859–76.
Franchi, Stefano. "Chess, Games, and Flies." *Essays in Philosophy* 6 (2005), 1–36.
Freedberg, David. *The Power of Images: Studies in the History and Theory of Response*. Chicago: University of Chicago Press, 1989.
Friedman, Ted. "Making Sense of Software: Computer Games and Interactive Textuality." In *Cybersociety: Computer-Mediated Communication and Community*, edited by Steve Jones (Thousand Oaks, CA: Sage, 1995), 73–89.
Gadamer, Hans-Georg. *Truth and Method*. London: Sheed and Ward, 1975.
Gallagher, Rob. *Videogames, Identity and Digital Subjectivity*. London: Routledge, 2017.
Galloway, Alexander R. *Gaming: Essays on Algorithmic Culture*. Minneapolis: University of Minnesota Press, 2006.
Galloway, Alexander R. *The Interface Effect*. New York: Polity Press, 2012.
Gandy, Robin. "Human versus Mechanical Intelligence." In *Machines and Thought: The Legacy of Alan Turing*, edited by Peter Millican and Andy Clark (New York: Clarendon Press, 1999), 125–36.
Garfinkel, Simson. *Architects of the Information Society: 35 Years of the Laboratory for Computer Science at MIT*. Cambridge, MA: MIT Press, 1999.
Gehl, Robert W. and Maria Bakardjieva, eds. *Socialbots and Their Friends: Digital Media and the Automation of Sociality*. London: Routledge, 2018.
Gell, Alfred. *Art and Agency: An Anthropological Theory*. Oxford: Clarendon Press, 1998.
Geller, Tom. "Overcoming the Uncanny Valley." *IEEE Computer Graphics and Applications* 28.4 (2008), 11–17.
Geoghegan, Bernard Dionysius. "Agents of History: Autonomous Agents and Crypto-Intelligence." *Interaction Studies* 9 (2008): 403–14.
Geoghegan, Bernard Dionysius. "The Cybernetic Apparatus: Media, Liberalism, and the Reform of the Human Sciences." PhD diss., Northwestern University, 2012.
Geoghegan, Bernard Dionysius. "Visionäre Informatik: Notizen über Vorführungen von Automaten und Computern, 1769–1962." *Jahrbuch für Historische Bildungsforschung* 20 (2015), 177–98.
Giappone, Krista Bonello Rutter. "Self-Reflexivity and Humor in Adventure Games." *Game Studies* 15.1 (2015). Available at http://gamestudies.org/1501/articles/bonello_k. Retrieved 7 January 2020.
Giddens, Anthony. *The Consequences of Modernity*. London: Wiley, 2013.
Gillmor, Dan. "Bubba Meets Microsoft: Bob, You Ain't Gonna Like This." *San Jose Mercury News*, 6 May 1995, 1D.

Gitelman, Lisa. *Always Already New: Media, History and the Data of Culture*. Cambridge, MA: MIT Press, 2006.
Gitelman, Lisa. *Paper Knowledge: Toward a Media History of Documents*. Durham, NC: Duke University Press, 2014.
Gitelman, Lisa. *Scripts, Grooves, and Writing Machines: Representing Technology in the Edison Era*. Stanford, CA: Stanford University Press, 1999.
Goffman, Erving. *Frame Analysis: An Essay on the Organization of Experience*. Cambridge, MA: Harvard University Press, 1974.
Goldman, Eric. "Search Engine Bias and the Demise of Search Engine Utopianism." In *Web Search: Multidisciplinary Perspectives*, edited by Amanda Spink and Michael Zimmer (Berlin: Springer, 2008), 121–33.
Golumbia, David. *The Cultural Logic of Computation*. Cambridge, MA: Harvard University Press, 2009.
Gombrich, Ernst Hans. *Art and Illusion: A Study in the Psychology of Pictorial Representation*. London: Phaidon, 1977.
Gong, Li, and Clifford Nass. "When a Talking-Face Computer Agent Is Half-human and Half-humanoid: Human Identity and Consistency Preference." *Human Communication Research* 33.2 (2007), 163–93.
Gooday, Graeme. "Re-writing the 'Book of Blots': Critical Reflections on Histories of Technological 'Failure.'" *History and Technology* 14 (1998), 265–91.
Goode, Luke. "Life, but Not as We Know It: AI and the Popular Imagination." *Culture Unbound: Journal of Current Cultural Research* 10 (2018), 185–207.
Goodfellow, Ian, Yoshua Bengio, and Aaron Courville. *Deep Learning*. Cambridge, MA: MIT Press, 2016.
"Google Assistant." Google, 2019. Available at https://assistant.google.com/. Retrieved 12 December 2019.
Granström, Helena, and Bo Göranzon. "Turing's Man: A Dialogue." *AI & Society* 28.1 (2013), 21–25.
Grau, Oliver. *Virtual Art: From Illusion to Immersion*. Cambridge, MA: MIT Press, 2003.
Greenberger, Martin. "The Two Sides of Time Sharing." Working paper, Sloan School of Management and Project MAC, Massachusetts Institute of Technology, 1965.
Greenfield, Adam. *Radical Technologies: The Design of Everyday Life*. New York: Verso, 2017.
Grudin, Jonathan. "The Computer Reaches Out: The Historical Continuity of Interface Design." In *Proceedings of the SIGCHI Conference on Human Factors in Computing Systems* (Chicago: ACM, 1990), 261–68.
Grudin, Jonathan. "Turing Maturing: The Separation of Artificial Intelligence and Human-Computer Interaction." *Interactions* 13 (2006), 54–57.
Gunkel, David J. "Communication and Artificial Intelligence: Opportunities and Challenges for the 21st Century." *Communication+1* 1.1 (2012): 1–25.
Gunkel, David J. *Gaming the System: Deconstructing Video Games, Games Studies, and Virtual Worlds*. Bloomington: Indiana University Press, 2018.
Gunkel, David J. *An Introduction to Communication and Artificial Intelligence*. Cambridge: Polity Press, 2020.
Gunkel, David J. *The Machine Question: Critical Perspectives on AI, Robots, and Ethics*. MIT Press, 2012.
Gunkel, David J. "Other Things: AI, Robots, and Society." In *A Networked Self and Human Augmentics, Artificial Intelligence, Sentience*, edited by Zizi Papacharissi (New York: Routledge, 2018), 51–68.

Gunkel, David J. *Robot Rights*. Cambridge, MA: MIT Press, 2018.
Gunkel, David J. "Second Thoughts: Toward a Critique of the Digital Divide." *New Media & Society* 5.4 (2003), 499–522.
Gunning, Tom. "An Aesthetic of Astonishment: Early Film and the (In)credulous Spectator." *Art and Text* 34 (1989), 31–45.
Gunning, Tom. "Heard over the Phone: The Lonely Villa and the de Lorde Tradition of the Terrors of Technology." *Screen* 32.2 (1991), 184–96.
Guzman, Andrea L. "Beyond Extraordinary: Theorizing Artificial Intelligence and the Self in Daily Life." In *A Networked Self and Human Augmentics, Artificial Intelligence, Sentience*, edited by Zizi Papacharissi (New York: Routledge, 2018), 83–96.
Guzman, Andrea L., ed. *Human-Machine Communication: Rethinking Communication, Technology, and Ourselves*. New York: Peter Lang, 2018.
Guzman, Andrea L. "Imagining the Voice in the Machine: The Ontology of Digital Social Agents." PhD diss., University of Illinois at Chicago, 2015.
Guzman, Andrea L. "Making AI Safe for Humans: A Conversation with Siri." In *Socialbots and Their Friends: Digital Media and the Automation of Sociality*, edited by Robert W. Gehl and Maria Bakardjieva (London: Routledge, 2017), 69–85.
Guzman, Andrea L. "The Messages of Mute Machines: Human-Machine Communication with Industrial Technologies." *Communication+1* 5.1 (2016), 1–30.
Guzman, Andrea L. "Voices in and of the Machine: Source Orientation toward Mobile Virtual Assistants." *Computers in Human Behavior* 90 (2019), 343–50.
Guzman, Andrea L., and Seth C. Lewis. "Artificial Intelligence and Communication: A Human–Machine Communication Research Agenda." *New Media & Society* 22.1 (2020), 70–86.
Haken, Hermann, Anders Karlqvist, and Uno Svedin, eds. *The Machine as Metaphor and Tool*. Berlin: Springer, 1993.
Hall, Stuart, ed. *Representation: Cultural Representations and Signifying Practices*. London: Sage, 1997.
Ham, Jaap, Raymond H. Cuijpers, and John-John Cabibihan. "Combining Robotic Persuasive Strategies: The Persuasive Power of a Storytelling Robot That Uses Gazing and Gestures." *International Journal of Social Robotics* 7 (2015), 479–87.
Harnad, Stevan. "The Turing Test Is Not a Trick: Turing Indistinguishability Is a Scientific Criterion." *SIGART Bulletin* 3.4 (1992), 9–10.
Hayes, Joy Elizabeth, and Kathleen Battles. "Exchange and Interconnection in US Network Radio: A Reinterpretation of the 1938 War of the Worlds Broadcast." *Radio Journal: International Studies in Broadcast & Audio Media* 9.1 (2011), 51–62.
Hayles, N. Katherine. *How We Became Posthuman: Virtual Bodies in Cybernetics, Literature, and Informatics*. Chicago: University of Chicago Press, 1999.
Hayles, N. Katherine. *How We Think: Digital Media and Contemporary Technogenesis*. Chicago: University of Chicago Press, 2012.
Hayles, N. Katherine. *Writing Machines*. Cambridge, MA: MIT Press, 2002.
Haugeland, John. *Artificial Intelligence: The Very Idea*. Cambridge, MA: MIT Press, 1985.
Heidorn, George E. "English as a Very High Level Language for Simulation Programming." *ACM SIGPLAN Notices* 9.4 (1974), 91–100.

Heyer, Paul. "America under Attack I: A Reassessment of Orson Welles' 1938 War of the Worlds Broadcast." *Canadian Journal of Communication* 28.2 (2003), 149–66.
Henrickson, Leah. "Computer-Generated Fiction in a Literary Lineage: Breaking the Hermeneutic Contract." *Logos* 29.2–3 (2018), 54–63.
Henrickson, Leah. "Tool vs. Agent: Attributing Agency to Natural Language Generation Systems." *Digital Creativity* 29.2–3 (2018): 182–90.
Henrickson, Leah. "Towards a New Sociology of the Text: The Hermeneutics of Algorithmic Authorship." PhD diss., Loughborough University, 2019.
Hepp, Andreas. "Artificial Companions, Social Bots and Work Bots: Communicative Robots as Research Objects of Media and Communication Studies." *Media, Culture and Society* 42.7-8 (2020), 1410–26.
Hepp, Andreas. *Deep Mediatization.* London: Routledge, 2019.
Hester, Helen. "Technically Female: Women, Machines, and Hyperemployment." *Salvage* 3 (2016). Available at https://salvage.zone/in-print/technically-female-women-machines-and-hyperemployment. Retrieved 30 December 2019.
Hicks, Marie. *Programmed Inequality: How Britain Discarded Women Technologists and Lost Its Edge in Computing.* Cambridge, MA: MIT Press, 2017.
Highmore, Ben. "Machinic Magic: IBM at the 1964–1965 New York World's Fair." *New Formations* 51.1 (2003), 128–48.
Hill, David W. "The Injuries of Platform Logistics." *Media, Culture & Society*, published online before print 21 July 2019, doi: 0163443719861840.
Hjarvard, Stig. "The Mediatisation of Religion: Theorising Religion, Media and Social Change." *Culture and Religion* 12.2 (2011), 119–35.
Hofer, Margaret K. *The Games We Played: The Golden Age of Board and Table Games.* New York: Princeton Architectural Press, 2003.
Hoffman, Donald D. *The Case against Reality: How Evolution Hid the Truth from Our Eyes.* London: Penguin, 2019.
Hollings, Christopher, Ursula Martin, and Adrian Rice. *Ada Lovelace: The Making of a Computer Scientist.* Oxford: Bodleian Library, 2018.
Holtgraves, T. M., Stephen J. Ross, C. R. Weywadt, and T. L. Han. "Perceiving Artificial Social Agents." *Computers in Human Behavior* 23 (2007), 2163–74.
Hookway, Branden. *Interface.* Cambridge, MA: MIT Press, 2014.
Hoy, Matthew B. "Alexa, Siri, Cortana, and More: An Introduction to Voice Assistants." *Medical Reference Services Quarterly* 37.1 (2018), 81–88.
Hu, Tung-Hui. *A Prehistory of the Cloud.* Cambridge, MA: MIT Press, 2015.
Huhtamo, Erkki. "Elephans Photographicus: Media Archaeology and the History of Photography." In *Photography and Other Media in the Nineteenth Century*, edited by Nicoletta Leonardi and Simone Natale (University Park: Penn State University Press, 2018), 15–35.
Huhtamo, Erkki. *Illusions in Motion: Media Archaeology of the Moving Panorama and Related Spectacles.* Cambridge, MA: MIT Press, 2013.
Humphry, Justine, and Chris Chesher. "Preparing for Smart Voice Assistants: Cultural Histories and Media Innovations." *New Media and Society*, published online before print 22 May 2020, doi: 10.1177/1461444820923679.
Humphrys, Mark. "How My Program Passed the Turing Test." In *Parsing the Turing Test*, edited by Robert Epstein, Gary Roberts, and Grace Beber (Amsterdam: Springer), 237–60.

Hutchens, Jason L. "How to Pass the Turing Test by Cheating." Research report. School of Electrical, Electronic and Computer Engineering, University of Western Australia, Perth, 1997.

Huizinga, Johan. *Homo Ludens: A Study of the Play Element in Culture*. London: Maurice Temple Smith, 1970.

Hwang, Tim, Ian Pearce, and Max Nanis. "Socialbots: Voices from the Fronts." *Interactions* 19.2 (2012), 38–45.

Hyman, R. "The Psychology of Deception." *Annual Review of Psychology* 40 (1989), 133–54.

Idone Cassone, Vincenzo and Mattia Thibault. "I Play, Therefore I Believe." In *Believing in Bits: Digital Media and the Supernatural*, edited by Simone Natale and Diana W. Pasulka (Oxford: Oxford University Press, 2019), 73–90.

"I Think, Therefore I'm RAM." *Daily Telegraph*, 26 December 1997, 14.

Jastrow, Joseph. *Fact and Fable in Psychology*. Boston: Houghton Mifflin, 1900.

Jerz, Dennis G. "Somewhere Nearby Is Colossal Cave: Examining Will Crowther's Original 'Adventure' in Code and in Kentucky." *Digital Humanities Quarterly* 1.2 (2007). Available at http://www.digitalhumanities.org/dhq/vol/1/2/000009/000009.html. Retrieved 7 January 2020.

Johnson, Steven. *Wonderland: How Play Made the Modern World*. London: Pan Macmillan, 2016.

Johnston, John. *The Allure of Machinic Life: Cybernetics, Artificial Life, and the New AI*. Cambridge, MA: MIT Press, 2008.

Jones, Paul. "The Technology Is Not the Cultural Form? Raymond Williams's Sociological Critique of Marshall McLuhan." *Canadian Journal of Communication* 23.4 (1998), 423–46.

Jones, Steve. "How I Learned to Stop Worrying and Love the Bots." *Social Media and Society* 1 (2015), 1–2.

Jørgensen, Kristine. *Gameworld Interfaces*. Cambridge, MA: MIT Press, 2013.

Juul, Jesper. *Half-Real: Video Games between Real Rules and Fictional Worlds*. Cambridge, MA: MIT Press, 2011.

Karppi, Tero. *Disconnect: Facebook's Affective Bonds*. Minneapolis: University of Minnesota Press, 2018.

Katzenbach, Christian, and Lena Ulbricht. "Algorithmic governance." *Internet Policy Review* 8.4 (2019). Available at https://policyreview.info/concepts/algorithmic-governance. Retrieved 10 November 2020.

Kelion, Leo. "Amazon Alexa Gets Samuel L Jackson and Celebrity Voices." *BBC News*, 25 September 2019. Available at https://www.bbc.co.uk/news/technology-49829391. Retrieved 12 December 2019.

Kelleher, John D. *Deep Learning*. Cambridge, MA: MIT Press, 2019.

Kim, Youjeong, and S. Shyam Sundar. "Anthropomorphism of Computers: Is It Mindful or Mindless?" *Computers in Human Behavior* 28.1 (2012), 241–50.

King, William Joseph. "Anthropomorphic Agents: Friend, Foe, or Folly" *HITL Technical Memorandum* M-95-1 (1995). Avalailable at http://citeseerx.ist.psu.edu/viewdoc/download?doi=10.1.1.57.3474&rep=rep1&type=pdf. Retrieved 10 November 2020.

Kirschenbaum, Matthew G. *Mechanisms: New Media and the Forensic Imagination*. Cambridge, MA: MIT Press, 2008.

Kittler, Friedrich. *Gramophone, Film, Typewriter*. Stanford, CA: Stanford University Press, 1999.

Kline, Ronald R. "Cybernetics, Automata Studies, and the Dartmouth Conference on Artificial Intelligence." *IEEE Annals of the History of Computing* 33 (2011), 5–16.
Kline, Ronald R. *The Cybernetics Moment: Or Why We Call Our Age the Information Age*. Baltimore: John Hopkins University Press, 2015.
Kohler, Robert. *Lords of the Fly: Drosophila Genetics and the Experimental Life*. Chicago: University of Chicago Press, 1994.
Kocurek, Carly A. "The Agony and the Exidy: A History of Video Game Violence and the Legacy of Death Race." *Game Studies* 12.1 (2012). Available at http://gamestudies.org/1201/articles/carly_kocurek. Retrieved 10 February 2020.
Korn, James H. *Illusions of Reality: A History of Deception in Social Psychology*. Albany: State University of New York Press, 1997.
Krajewski, Markus. *The Server: A Media History from the Present to the Baroque*. New Haven, CT: Yale University Press, 2018.
Kris, Ernst and Otto Kurz. *Legend, Myth, and Magic in the Image of the Artist: A Historical Experiment*. New Haven, CT: Yale University Press, 1979.
Kurzweil, Ray. *The Singularity Is Near: When Humans Transcend Biology*. London: Penguin, 2005.
Laing, Dave. "A Voice without a Face: Popular Music and the Phonograph in the 1890s." *Popular Music* 10.1 (1991), 1–9.
Lakoff, George, and Mark Johnson. *Metaphor We Live By*. Chicago: University of Chicago Press, 1980.
Lamont, Peter. *Extraordinary Beliefs: A Historical Approach to a Psychological Problem*. Cambridge: Cambridge University Press, 2013.
Langer, Ellen J. "Matters of Mind: Mindfulness/Mindlessness in Perspective." *Consciousness and Cognition* 1.3 (1992), 289–305.
Lanier, Jaron. *You Are Not a Gadget*. London: Penguin Books, 2011).
Lankoski, Petri. "Player Character Engagement in Computer Games." *Games and Culture* 6 (2011), 291–311.
Latour, Bruno. *Pandora's Hope: Essays on the Reality of Science Studies*. Cambridge, MA: Harvard University Press, 1999.
Latour, Bruno. *The Pasteurization of France*. Cambridge, MA: Harvard University Press, 1993.
Latour, Bruno. *We Have Never Been Modern*. Cambridge, MA: Harvard University Press, 1993.
Laurel, Brenda. *Computers as Theatre*. Upper Saddle River, NJ: Addison-Wesley, 2013.
Laurel, Brenda. "Interface Agents: Metaphors with Character." In *Human Values and the Design of Computer Technology*, edited by Batya Friedman (Stanford, CA: CSLI, 1997), 207–19.
Lee, Kwan Min. "Presence, Explicated." *Communication Theory* 14.1 (2004), 27–50.
Leeder, Murray. *The Modern Supernatural and the Beginnings of Cinema*. Basingstoke, UK: Palgrave Macmillan, 2017.
Leff, Harvey S., and Andrew F. Rex, eds. *Maxwell's Demon: Entropy, Information, Computing*. Princeton, NJ: Princeton University Press, 2014.
Leja, Michael. *Looking Askance: Skepticism and American Art from Eakins to Duchamp*. Berkeley: University of California Press, 2004.
Leonard, Andrew. *Bots: The Origin of a New Species*. San Francisco: HardWired, 1997.
Lesage, Frédérik. "A Cultural Biography of Application Software." In *Advancing Media Production Research: Shifting Sites, Methods, and Politics*, edited by Chris Paterson, D. Lee, A. Saha, and A. Zoellner (London: Palgrave, 2015), 217–32.

Lesage, Frédérik. "Popular Digital Imaging: Photoshop as Middlebroware." In *Materiality and Popular Culture*, edited by Anna Malinowska and Karolina Lebek (London: Routledge, 2016), 88–99.

Lesage, Frederik, and Simone Natale. "Rethinking the Distinctions between Old and New Media: Introduction." *Convergence* 25.4 (2019), 575–89.

Leslie, Ian. "Why Donald Trump Is the First Chatbot President." *New Statesman*, 13 November 2017. Available at https://www.newstatesman.com/world/north-america/2017/11/why-donald-trump-first-chatbot-president. Retrieved 27 November 2019.

Lessard, Jonathan. "Adventure before Adventure Games: A New Look at Crowther and Woods's Seminal Program." *Games and Culture* 8 (2013), 119–35.

Lessard, Jonathan, and Dominic Arsenault. "The Character as Subjective Interface." In *International Conference on Interactive Digital Storytelling* (Cham, Switzerland: Springer, 2016), 317–324.

Lester, James C., S. Todd Barlow, Sharolyn A. Converse, Brian A. Stone, Susan E. Kahler, and Ravinder S. Bhogal. "Persona Effect: Affective Impact of Animated Pedagogical Agents." *CHI '97: Proceedings of the ACM SIGCHI Conference on Human Factors in Computing Systems* (1997), 359–66.

Levesque, Hector J. *Common Sense, the Turing Test, and the Quest for Real AI: Reflections on Natural and Artificial Intelligence*. Cambridge, MA: MIT Press, 2017.

Licklider, Joseph C. R. "Man-Computer Symbiosis." *IRE Transactions on Human Factors in Electronics* HFE-1 (1960): 4–11.

Licklider, Joseph C. R., and Robert W. Taylor. "The Computer as a Communication Device." *Science and Technology* 2 (1968), 2–5.

Liddy, Elizabeth D. "Natural Language Processing." In *Encyclopedia of Library and Information Science* (New York: Marcel Decker, 2001). Available at https://surface.syr.edu/cgi/viewcontent.cgi?filename=0&article=1043&context=istpub&type=additional. Retrieved 8 November 2020.

Lighthill, James. "Artificial Intelligence: A General Survey." In *Artificial Intelligence: A Paper Symposium* (London: Science Research Council, 1973), 1–21.

Lindquist, Christopher. "Quest for Machines That Think." *Computerworld*, 18 November 1991, 22.

Lipartito, Kenneth. "Picturephone and the Information Age: The Social Meaning of Failure." *Technology and Culture* 44 (2003), 50–81.

Lippmann, Walter. *Public Opinion*. New York: Harcourt, Brace, 1922.

Littlefield, Melissa M. *The Lying Brain: Lie Detection in Science and Science Fiction*. Ann Arbor: University of Michigan Press, 2011.

Liu, Lydia H. *The Freudian Robot: Digital Media and the Future of the Unconscious*. Chicago: University of Chicago Press, 2010.

Zachary, Loeb. "Introduction." In Joseph Weizenbaum, *Islands in the Cyberstream: Seeking Havens of Reason in a Programmed Society* (Sacramento, CA: Liewing Books, 2015), 1–25.

Loebner, Hugh. "The Turing Test." *New Atlantis* 12 (2006), 5–7.

Loiperdinger, Martin. "Lumière's Arrival of the Train: Cinema's Founding Myth." *Moving Image* 4.1 (2004), 89–118.

Lomborg, Stine, and Patrick Heiberg Kapsch. "Decoding Algorithms." *Media, Culture & Society* 42.5 (2019), 745–61.

Luger, George F., and Chayan Chakrabarti. "From Alan Turing to Modern AI: Practical Solutions and an Implicit Epistemic Stance." *AI & Society* 32.3 (2017), 321–38.

Luka Inc. "Replika." Available at https://replika.ai/. Retrieved 30 December 2019.
Łupkowski, Paweł, and Aleksandra Rybacka. "Non-cooperative Strategies of Players in the Loebner Contest." *Organon F* 23.3 (2016), 324–65.
MacArthur, Emily. "The iPhone Erfahrung: Siri, the Auditory Unconscious, and Walter Benjamin's Aura." In *Design, Mediation, and the Posthuman*, edited by Dennis M. Weiss, Amy D. Propen, and Colbey Emmerson Reid (Lanham, MD: Lexington Books, 2014), 113–27.
Mackenzie, Adrian. "The Performativity of Code Software and Cultures of Circulation." *Theory, Culture & Society* 22 (2005), 71–92.
Mackinnon, Lee. "Artificial Stupidity and the End of Men." *Third Text* 31.5–6 (2017), 603–17.
Magid, Lawrence. "Microsoft Bob: No Second Chance to Make a First Impression." *Washington Post*, 16 January 1995, F18.
Mahon, James Edwin. "The Definition of Lying and Deception." In *The Stanford Encyclopedia of Philosophy*, edited by Edward N. Zalta. 2015. Available at https://plato.stanford.edu/archives/win2016/entries/lying-definition/. Retrieved 15 July 2020.
Mahoney, Michael S. "What Makes the History of Software Hard." *IEEE Annals of the History of Computing* 30.3 (2008), 8–18.
Malin, Brenton J. *Feeling Mediated: A History of Media Technology and Emotion in America*. New York: New York University Press, 2014.
Manes, Stephen. "Bob: Your New Best Friend's Personality Quirks." *New York Times*, 17 January 1995, C8.
Manning, Christopher D., Prabhakar Raghavan, and Hinrich Schütze. *Introduction to Information Retrieval*. Cambridge: Cambridge University Press, 2008.
Manon, Hugh S. "Seeing through Seeing through: The Trompe l'oeil Effect and Bodily Difference in the Cinema of Tod Browning." *Framework* 47.1 (2006), 60–82.
Manovich, Lev. "How to Follow Software Users." available at http://manovich.net/content/04-projects/075-how-to-follow-software-users/72_article_2012.pdf. Retrieved 10 February 2020.
Manovich, Lev. *The Language of New Media*. Cambridge, MA: MIT Press, 2002.
Marenko, Betti, and Philip Van Allen. "Animistic Design: How to Reimagine Digital Interaction between the Human and the Nonhuman." *Digital Creativity* 27.1 (2016): 52–70.
Marino, Mark C. "I, Chatbot: The Gender and Race Performativity of Conversational Agents." PhD diss., University of California Riverside, 2006.
Markoff, John. "Can Machines Think? Humans Match Wits." *New York Times*, 8 November 1991, 1.
Markoff, John. "So Who's Talking: Human or Machine?" *New York Times*, 5 November 1991, C1.
Markoff, John. "Theaters of High Tech." *New York Times*, 12 January 1992, 15.
Martin, Clancy W., ed. *The Philosophy of Deception*. Oxford: Oxford University Press, 2009.
Martin, C. Dianne. "The Myth of the Awesome Thinking Machine." *Communications of the ACM* 36 (1993): 120–33.
Mauldin, Michael L. "ChatterBots, TinyMuds, and the Turing Test: Entering the Loebner Prize Competition." *Proceedings of the National Conference on Artificial Intelligence* 1 (1994), 16–21.
McCarthy, John. "Information." *Scientific American* 215.3 (1966), 64–72.

McCorduck, Pamela. *Machines Who Think: A Personal Inquiry into the History and Prospects of Artificial Intelligence*. San Francisco: Freeman, 1979.
McCracken, Harry. "The Bob Chronicles." *Technologizer*, 29 March 2010. Available at https://www.technologizer.com/2010/03/29/microsoft-bob/. Retrieved 19 November 2019.
McCulloch, Warren, and Walter Pitts. "A Logical Calculus of the Ideas Immanent in Nervous Activity." *Bulletin of Mathematical Biology* 5 (1943): 115–33.
McKee, Heidi. *Professional Communication and Network Interaction: A Rhetorical and Ethical Approach*. London: Routledge, 2017.
McKelvey, Fenwick. *Internet Daemons: Digital Communications Possessed*. Minneapolis: University of Minnesota Press, 2018.
McLean, Graeme, and Ko Osei-frimpong. "Hey Alexa . . . Examine the Variables Influencing the Use of Artificial Intelligent In-home Voice Assistants." *Computers in Human Behavior* 99 (2019), 28–37.
McLuhan, Marshall. *Understanding Media: The Extensions of Man*. Toronto: McGraw-Hill, 1964.
Meadow, Charles T. *Man-Machine Communication*. New York: Wiley, 1970.
Messeri, Lisa, and Janet Vertesi. "The Greatest Missions Never Flown: Anticipatory Discourse and the Projectory in Technological Communities." *Technology and Culture* 56.1 (2015), 54–85.
"Microsoft Bob." Toastytech.com, Available at http://toastytech.com/guis/bob.html. Retrieved 19 November 2019.
"Microsoft Bob Comes Home: A Breakthrough in Home Computing." PR Newswire Association, 7 January 1995, 11:01 ET.
Miller, Kiri. "Grove Street Grimm: Grand Theft Auto and Digital Folklore." *Journal of American Folklore* 121.481 (2008), 255–85.
Mindell, David. *Between Human and Machine: Feedback, Control, and Computing before Cybernetics*. Baltimore: Johns Hopkins University Press, 2002.
Minsky, Marvin. "Artificial Intelligence." *Scientific American* 215 (1966), 246–60.
Minsky, Marvin. "Problems of Formulation for Artificial Intelligence." *Proceedings of Symposia in Applied Mathematics* 14 (1962), 35–46.
Minsky, Marvin, ed. *Semantic Information Processing*. Cambridge, MA: MIT Press, 1968.
Minsky, Marvin. *The Society of Mind*. New York: Simon and Schuster, 1986.
Minsky, Marvin. "Some Methods of Artificial Intelligence and Heuristic Programming." *Proceeding of the Symposium on the Mechanization of Thought Processes* 1 (1959), 3–25.
Minsky, Marvin. "Steps toward Artificial Intelligence." *Proceedings of the IRE* 49.1 (1961), 8–30.
Monroe, John Warne. *Laboratories of Faith: Mesmerism, Spiritism, and Occultism in Modern France*. Ithaca, NY: Cornell University Press, 2008.
Montfort, Nick. "Zork." In *Space Time Play*, edited by Friedrich von Borries, Steffen P. Walz, and Matthias Böttger (Basel, Switzerland: Birkhäuser, 2007), 64–65.
Moor, James H, ed. *The Turing Test: The Elusive Standard of Artificial Intelligence*. Dordrecht, Netherlands: Kluwer Academic, 2003.
Moore, Matthew. "Alexa, Why Are You a Bleeding-Heart Liberal?" *Times* (London), 12 December 2017. Available at https://www.thetimes.co.uk/article/8869551e-dea5-11e7-872d-4b5e82b139be. Retrieved 15 December 2019.
Moore, Phoebe. "The Mirror for (Artificial) Intelligence in Capitalism." *Comparative Labour Law and Policy Journal* 44.2 (2020), 191–200.

Moravec, Hans. *Mind Children: The Future of Robot and Human Intelligence*. Cambridge, MA: Harvard University Press, 1988.
Mori, Masahiro. "The Uncanny Valley." *IEEE Spectrum*, 12 June 2012. Available at https://spectrum.ieee.org/automaton/robotics/humanoids/the-uncanny-valley. Retrieved 9 November 2020.
Morus, Iwan Rhys. *Frankenstein's Children: Electricity, Exhibition, and Experiment in Early-Nineteenth-Century London*. Princeton, NJ: Princeton University Press, 1998.
Mosco, Vincent. *The Digital Sublime: Myth, Power, and Cyberspace*. Cambridge, MA: MIT Press, 2004.
Müggenburg, Jan. "Lebende Prototypen und lebhafte Artefakte. Die (Un-) Gewissheiten Der Bionik." *Ilinx—Berliner Beiträge Zur Kulturwissenschaft* 2 (2011), 1–21.
Muhle, Florian. "Embodied Conversational Agents as Social Actors? Sociological Considerations in the Change of Human-Machine Relations in Online Environments." In *Socialbots and Their Friends: Digital Media and the Automation of Sociality*, edited by Robert W. Gehl and Maria Bakardjeva (London: Routledge, 2017), 86–109.
Mühlhoff, Rainer. "Human-Aided Artificial Intelligence: Or, How to Run Large Computations in Human Brains? Toward a Media Sociology of Machine Learning." *New Media and Society*, published online before print 6 November 2019, doi: 10.1177/1461444819885334.
Münsterberg, Hugo. *American Problems from the Point of View of a Psychologist*. New York: Moffat, 1910.
Münsterberg, Hugo. *The Film: A Psychological Study*. New York: Dover, 1970.
Murray, Janet H. *Hamlet on the Holodeck: The Future of Narrative in Cyberspace*. Cambridge, MA: MIT Press, 1998.
Musès, Charles, ed. *Aspects of the Theory of Artificial Intelligence: Proceedings*. New York: Plenum Press, 1962.
Nadis, Fred. *Wonder Shows: Performing Science, Magic, and Religion in America*. New Brunswick, NJ: Rutgers University Press, 2005.
Nagel, Thomas. "What Is It Like to Be a Bat?" *Philosophical Review* 83.4 (1974), 435–50.
Nagy, Peter, and Gina Neff. "Imagined Affordance: Reconstructing a Keyword for Communication Theory." *Social Media and Society* 1.2 (2015), doi: 10.1177/2056305115603385.
Nass, Clifford, and Scott Brave. *Wired for Speech: How Voice Activates and Advances the Human-Computer Relationship*. Cambridge, MA: MIT Press, 2005.
Nass, Clifford, and Youngme Moon. "Machines and Mindlessness: Social Responses to Computers." *Journal of Social Issues* 56.1 (2000), 81–103.
Natale, Simone. "All That's Liquid." *New Formations* 91 (2017), 121–23.
Natale, Simone. "Amazon Can Read Your Mind: A Media Archaeology of the Algorithmic Imaginary." In *Believing in Bits: Digital Media and the Supernatural*, edited by Simone Natale and Diana Pasulka (Oxford: Oxford University Press, 2019), 19–36.
Natale, Simone. "The Cinema of Exposure: Spiritualist Exposés, Technology, and the Dispositif of Early Cinema." *Recherches Sémiotiques/Semiotic Inquiry* 31.1 (2011), 101–17.
Natale, Simone. "Communicating through or Communicating with: Approaching Artificial Intelligence from a Communication and Media Studies Perspective."

Communication Theory, published online before print 24 September 2020. Available at https://doi.org/10.1093/ct/qtaa022. Retrieved 11 November 2020.
Natale, Simone. "If Software Is Narrative: Joseph Weizenbaum, Artificial Intelligence and the Biographies of ELIZA." *New Media & Society* 21.3 (2018), 712–28.
Natale, Simone. "Introduction: New Media and the Imagination of the Future." *Wi: Journal of Mobile Media* 8.2 (2014), 1–8.
Natale, Simone. *Supernatural Entertainments: Victorian Spiritualism and the Rise of Modern Media Culture*. University Park: Penn State University Press, 2016.
Natale, Simone. "Unveiling the Biographies of Media: On the Role of Narratives, Anecdotes and Storytelling in the Construction of New Media's Histories." *Communication Theory* 26.4 (2016), 431–49.
Natale, Simone. "Vinyl Won't Save Us: Reframing Disconnection as Engagement." *Media, Culture and Society* 42.4 (2020), 626–33.
Natale, Simone, and Andrea Ballatore. "Imagining the Thinking Machine: Technological Myths and the Rise of Artificial Intelligence." *Convergence: The International Journal of Research into New Media Technologies* 26 (2020), 3–18.
Natale, Simone, Paolo Bory, and Gabriele Balbi. "The Rise of Corporational Determinism: Digital Media Corporations and Narratives of Media Change." *Critical Studies in Media Communication* 36.4 (2019), 323–38.
Natale, Simone, and Diana W. Pasulka, eds. *Believing in Bits: Digital Media and the Supernatural*. Oxford: Oxford University Press, 2019.
Neff, Gina, and Peter Nagy. "Talking to Bots: Symbiotic Agency and the Case of Tay." *International Journal of Communication* 10 (2016), 4915–31.
Neudert, Lisa-Maria. "Future Elections May Be Swayed by Intelligent, Weaponized Chatbots." *MIT Technology Review* 121.5 (2018), 72–73.
Newell, Allen, John Calman Shaw, and Herbert A. Simon. "Chess-Playing Programs and the Problem of Complexity." *IBM Journal of Research and Development* 2 (1958): 320–35.
Newton, Julianne H. "Media Ecology." In *The International Encyclopedia of Communication*, edited by W. Donsbach (London: Wiley, 2015), 1–5.
Nickerson, Raymond S., Jerome I. Elkind, and Jaime R. Carbonell. "Human Factors and the Design of Time Sharing Computer Systems." *Human Factors* 10.2 (1968), 127–33.
Niculescu, Andreea, Betsy van Dijk, Anton Nijholt, Haizhou Li, and Swee Lan See. "Making Social Robots More Attractive: The Effects of Voice Pitch, Humor and Empathy." *International Journal of Social Robotics* 5.2 (2013), 171–91.
Nilsson, Nils J. *The Quest for Artificial Intelligence*. Cambridge: Cambridge University Press, 2013.
Nishimura, Keiko. "Semi-autonomous Fan Fiction: Japanese Character Bots and Non-human Affect." In *Socialbots and Their Friends: Digital Media and the Automation of Sociality*, edited by Robert W. Gehl and Maria Bakardjieva (London: Routledge, 2018), 128–44.
Noakes, Richard J. "Telegraphy Is an Occult Art: Cromwell Fleetwood Varley and the Diffusion of Electricity to the Other World." *British Journal for the History of Science* 32.4 (1999), 421–59.
Noble, Safiya Umoja. *Algorithms of Oppression: How Search Engines Reinforce Racism*. New York: New York University Press, 2018.
Norman, Donald A. *Emotional Design: Why We Love (or Hate) Everyday Things*. New York: Basic Books, 2004.

North, Dan. "Magic and Illusion in Early Cinema." *Studies in French Cinema* 1 (2001), 70–79.

"Number of Digital Voice Assistants in Use Worldwide from 2019 to 2023." *Statista*, 14 November 2019. Available at https://www.statista.com/statistics/973815/worldwide-digital-voice-assistant-in-use/. Retrieved 10 February 2020.

Oettinger, Anthony G. "The Uses of Computers in Science." *Scientific American* 215.3 (1966), 160–72.

O'Leary, Daniel E. "Google's Duplex: Pretending to Be Human." *Intelligent Systems in Accounting, Finance and Management* 26.1 (2019), 46–53.

Olson, Christi, and Kelly Kemery. "From Answers to Action: Customer Adoption of Voice Technology and Digital Assistants." *Microsoft Voice Report*, 2019. Available at https://about.ads.microsoft.com/en-us/insights/2019-voice-report. Retrieved 20 December 2019.

Ortoleva, Peppino. *Mediastoria*. Milan: Net, 2002.

Ortoleva, Peppino. *Miti a bassa intensità*. Turin: Einaudi, 2019.

Ortoleva, Peppin. "Modern Mythologies, the Media and the Social Presence of Technology." *Observatorio (OBS) Journal*, 3 (2009), 1–12.

Ortoleva, Peppino. "Vite Geniali: Sulle biografie aneddotiche degli inventori." *Intersezioni* 1 (1996), 41–61.

Papacharissi, Zizi, ed. *A Networked Self and Human Augmentics, Artificial Intelligence, Sentience*. New York: Routledge, 2019.

Parikka, Jussi. *What Is Media Archaeology?* Cambridge: Polity Press, 2012.

Parisi, David. *Archaeologies of Touch: Interfacing with Haptics from Electricity to Computing*. Minneapolis: University of Minnesota Press, 2018.

Park, David W., Nick Jankowski, and Steve Jones, eds. *The Long History of New Media: Technology, Historiography, and Contextualizing Newness*. New York: Peter Lang, 2011.

Pask, Gordon. "A Discussion of Artificial Intelligence and Self-Organization." *Advances in Computers* 5 (1964), 109–226.

Peters, Benjamin. *How Not to Network a Nation: The Uneasy History of the Soviet Internet*. Cambridge, MA: MIT Press, 2016.

Peters, John Durham. *The Marvelous Cloud: Towards a Philosophy of Elemental Media*. Chicago: University of Chicago Press, 2015.

Peters, John Durham. *Speaking into the Air: A History of the Idea of Communication*. Chicago: University of Chicago Press, 1999.

Pettit, Michael. *The Science of Deception: Psychology and Commerce in America*. Chicago: University of Chicago Press, 2013.

Phan, Thao. "The Materiality of the Digital and the Gendered Voice of Siri." *Transformations* 29 (2017), 23–33.

Picard, Rosalind W. *Affective Computing*. Cambridge, MA: MIT Press, 2000.

Picker, John M. "The Victorian Aura of the Recorded Voice." *New Literary History* 32.3 (2001), 769–86.

Pickering, Michael. *Stereotyping: The Politics of Representation*. Basingstoke, UK: Palgrave, 2001.

Pieraccini, Roberto. *The Voice in the Machine: Building Computers That Understand Speech*. Cambridge, MA: MIT Press, 2012.

Poe, Edgar Allan. *The Raven; with, The Philosophy of Composition*. Wakefield, RI: Moyer Bell, 1996.

Pollini, Alessandro. "A Theoretical Perspective on Social Agency." *AI & Society* 24.2 (2009), 165–71.

Pooley, Jefferson, and Michael J. Socolow. "War of the Words: The Invasion from Mars and Its Legacy for Mass Communication Scholarship." In *War of the Worlds to Social Media: Mediated Communication in Times of Crisis*, edited by Joy Hayes, Kathleen Battles, and Wendy Hilton-Morrow (New York: Peter Lang, 2013), 35–56.

Porcheron, Martin, Joel E. Fischer, Stuart Reeves, and Sarah Sharples. "Voice Interfaces in Everyday Life." *CHI '18: Proceedings of the 2018 CHI Conference on Human Factors in Computing Systems* (2018), 1–12.

Powers, David M. W., and Christopher C. R. Turk. *Machine Learning of Natural Language*. London: Springer-Verlag, 1989.

Pruijt, Hans. "Social Interaction with Computers: An Interpretation of Weizenbaum's ELIZA and Her Heritage." *Social Science Computer Review* 24.4 (2006), 517–19.

Rabiner, Lawrence R., and Ronald W. Schafer. "Introduction to Digital Speech Processing." *Foundations and Trends in Signal Processing* 1.1–2 (2007), 1–194.

Rasskin-Gutman, Diego. *Chess Metaphors: Artificial Intelligence and the Human Mind*. Cambridge, MA: MIT Press, 2009.

Reeves, Byron, and Clifford Nass. *The Media Equation: How People Treat Computers, Television, and New Media like Real People and Places*. Stanford, CA: CSLI, 1996.

Rhee, Jennifer. "Beyond the Uncanny Valley: Masahiro Mori and Philip K. Dick's *Do Androids Dream of Electric Sheep*?" *Configurations* 21.3 (2013), 301–29.

Rhee, Jennifer. "Misidentification's Promise: The Turing Test in Weizenbaum, Powers, and Short." *Postmodern Culture* 20.3 (2010). Available online at https://muse.jhu.edu/article/444706. Retrieved 8 January 2020.

Riskin, Jessica. "The Defecating Duck, or, the Ambiguous Origins of Artificial Life." *Critical Inquiry* 29.4 (2003), 599–633.

Russell, Stuart J., and Peter Norvig. *Artificial Intelligence: A Modern Approach*. Upper Saddle River, NJ: Pearson Education, 2002.

Rutschmann, Ronja, and Alex Wiegmann. "No Need for an Intention to Deceive? Challenging the Traditional Definition of Lying." *Philosophical Psychology* 30.4 (2017), 438–57.

"Samuel L. Jackson—Celebrity Voice for Alexa." Amazon.com, N.d. Available at https://www.amazon.com/Samuel-L-Jackson-celebrity-voice/dp/B07WS3HN5Q. Retrieved 12 December 2019.

Samuel, Arthur L. "Some Studies in Machine Learning Using the Game of Checkers." *IBM Journal of Research and Development* 3 (1959), 210–29.

Saygin, Ayse Pinar, Ilyas Cicekli, and Varol Akman. "Turing Test: 50 Years Later." *Minds and Machines* 10 (2000), 463–518.

Schank, Roger C. *Tell Me a Story: Narrative and Intelligence*. Evanston, IL: Northwestern University Press, 1995.

Schank, Roger C., and Robert P. Abelson. *Scripts, Plans, Goals, and Understanding: An Inquiry into Human Knowledge Structures*. Hillsdale, NJ: Erlbaum, 1977.

Schiaffonati, Viola. *Robot, Computer ed Esperimenti*. Milano, Italy: Meltemi, 2020.

Schieber, Stuart, ed. *The Turing Test: Verbal Behavior as the Hallmark of Intelligence*. Cambridge, MA: MIT Press, 2003.

Scolari, Carlos A. *Las leyes de la interfaz*. Barcelona: Gedisa, 2018.

Sconce, Jeffrey. *The Technical Delusion: Electronics, Power, Insanity*. Durham, NC: Duke University Press, 2019.

Sconce, Jeffrey. *Haunted Media: Electronic Presence from Telegraphy to Television*. Durham, NC: Duke University Press, 2000.

Schuetzler, Ryan M., G. Mark Grimes, and Justin Scott Giboney. "The Effect of Conversational Agent Skill on User Behavior during Deception." *Computers in Human Behavior* 97 (2019), 250–59.
Schulte, Stephanie Ricker. *Cached: Decoding the Internet in Global Popular Culture*. New York: New York University Press, 2013.
Schüttpelz, Erhard. "Get the Message Through: From the Channel of Communication to the Message of the Medium (1945–1960)." In *Media, Culture, and Mediality. New Insights into the Current State of Research*, edited by Ludwig Jäger, Erika Linz, and Irmela Schneider (Bielefeld, Germany : Transcript, 2010), 109–38.
Searle, John R. "Minds, Brains, and Programs." *Behavioral and Brain Sciences* 3.3 (1980), 417–57.
Shannon, Claude. "The Mathematical Theory of Communication." In *The Mathematical Theory of Communication*, edited by Claude Elwood Shannon and Warren Weaver (Urbana: University of Illinois Press, 1949), 29–125.
Shaw, Bertrand. *Pygmalion*. New York: Brentano, 1916.
Shieber, Stuart, ed. *The Turing test: Verbal behavior as the hallmark of intelligence* (Cambridge, MA: MIT Press, 2004).
Shieber, Stuart. "Lessons from a Restricted Turing Test." *Communications of the Association for Computing Machinery* 37.6 (1994), 70–78.
Shrager, Jeff. "The Genealogy of Eliza." Elizagen.org, date unknown. Available at http://elizagen.org/. Retrieved 10 February 2020.
Siegel, Michael Steven. "Persuasive Robotics: How Robots Change Our Minds." PhD diss., Massachusetts Institute of Technology, 2009.
Simon, Bart. "Beyond Cyberspatial Flaneurie: On the Analytic Potential of Living with Digital Games." *Games and Culture* 1.1 (2006), 62–67.
Simon, Herbert. "Reflections on Time Sharing from a User's Point of View." *Computer Science Research Review* 45 (1966): 31–48.
Sirois-Trahan, Jean-Pierre. "Mythes et limites du train-qui-fonce-sur-les-spectateurs." In *Limina: Le Soglie Del Film*, edited by Veronica Innocenti and Valentina Re (Udine, Italy: Forum, 2004), 203–16.
Smith, Gary. *The AI Delusion*. Oxford: Oxford University Press, 2018.
Smith, Merritt Roe, and Leo Marx, eds. *Does Technology Drive History? The Dilemma of Technological Determinism*. Cambridge, MA: MIT Press, 1994.
Smith, Rebecca M. "Microsoft Bob to Have Little Steam, Analysts Say." *Computer Retail Week* 5.94 (1995), 37.
Sobchack, Vivian. "Science Fiction Film and the Technological Imagination." In *Technological Visions: The Hopes and Fears That Shape New Technologies*, edited by Marita Sturken, Douglas Thomas, and Sandra Ball-Rokeach (Philadelphia: Temple University Press, 2004), 145–58.
Solomon, Matthew. *Disappearing Tricks: Silent Film, Houdini, and the New Magic of the Twentieth Century*. Urbana: University of Illinois Press, 2010.
Solomon, Robert C. "Self, Deception, and Self-Deception in Philosophy." In *The Philosophy of Deception*, edited by Clancy W. Martin (Oxford: Oxford University Press, 2009), 15–36.
Soni, Jimmy, and Rob Goodman. *A Mind at Play: How Claude Shannon Invented the Information Age*. Simon and Schuster, 2017.
Sonnevend, Julia. *Stories without Borders: The Berlin Wall and the Making of a Global Iconic Event*. New York: Oxford University Press, 2016.
Sontag, Susan. *On Photography*. New York: Anchor Books, 1990.

Sproull, Lee, Mani Subramani, Sara Kiesler, Janet H. Walker, and Keith Waters. "When the Interface Is a Face." *Human–Computer Interaction* 11.2 (1996), 97–124.

Spufford, Francis, and Jennifer S. Uglow. *Cultural Babbage: Technology, Time and Invention*. London: Faber, 1996.

Stanyer, James, and Sabina Mihelj. "Taking Time Seriously? Theorizing and Researching Change in Communication and Media Studies." *Journal of Communication* 66.2 (2016), 266–79.

Steinel, Wolfgang, and Carsten KW De Dreu. "Social Motives and Strategic Misrepresentation in Social Decision Making." *Journal of Personality and Social Psychology* 86.3 (2004), 419–34.

Sterne, Jonathan. *The Audible Past: Cultural Origins of Sound Reproduction*. Durham, NC: Duke University Press, 2003.

Sterne, Jonathan. *MP3: The Meaning of a Format*. Durham, NC: Duke University Press, 2012.

Stokoe, Elizabeth, Rein Ove Sikveland, Saul Albert, Magnus Hamann, and William Housley. "Can Humans Simulate Talking Like Other Humans? Comparing Simulated Clients to Real Customers in Service Inquiries." *Discourse Studies* 22.1 (2020), 87–109.

Stork, David G., ed. *HAL's Legacy: 2001's Computer as Dream and Reality*. Cambridge, MA: MIT Press, 1997.

Streeter, Thomas. *The Net Effect: Romanticism, Capitalism, and the Internet*. New York: New York University Press, 2010.

Stroda, Una. "Siri, Tell Me a Joke: Is There Laughter in a Transhuman Future?" In *Spiritualities, Ethics, and Implications of Human Enhancement and Artificial Intelligence*, edited by Christopher Hrynkow (Wilmington, DE: Vernon Press, 2020), 69–85.

Suchman, Lucy. *Human-Machine Reconfigurations: Plans and Situated Actions*. Cambridge: Cambridge University Press, 2007.

Suchman, Lucy. *Plans and Situated Actions: The Problem of Human-Machine Communication*. Cambridge: Cambridge University Press, 1987.

Sussman, Mark. "Performing the Intelligent Machine: Deception and Enchantment in the Life of the Automaton Chess Player." *TDR/The Drama Review* 43.3 (1999), 81–96.

Sweeney, Miriam E. "Digital Assistants." In *Uncertain Archives: Critical Keywords for Big Data*, edited by Nanna Bonde Thylstrup, Daniela Agostinho, Annie Ring, Catherine D'Ignazio, and Kristin Veel (Cambridge, MA: MIT Press, 2020). Pre-print available at https://ir.ua.edu/handle/123456789/6348. Retrieved 7 November 2020.

Sweeney, Miriam E. "Not Just a Pretty (Inter)face: A Critical Analysis of Microsoft's 'Ms. Dewey.'" PhD diss., University of Illinois at Urbana-Champaign, 2013.

Tavinor, Grant. "Videogames and Interactive Fiction." *Philosophy and Literature* 29.1 (2005), 24–40.

Thibault, Ghislain. "The Automatization of Nikola Tesla: Thinking Invention in the Late Nineteenth Century." *Configurations* 21.1 (2013), 27–52.

Thorson, Kjerstin, and Chris Wells. "Curated Flows: A Framework for Mapping Media Exposure in the Digital Age." *Communication Theory* 26 (2016), 309–28.

Tognazzini, Bruce. "Principles, Techniques, and Ethics of Stage Magic and Their Application to Human Interface Design." In *CHI '93" Proceedings of the*

INTERACT '93 and CHI '93 Conference on Human Factors in Computing Systems (1993), 355–62.

Torrance, Thomas F. *The Christian Doctrine of God, One Being Three Persons*. London: Bloomsbury, 2016.

Towns, Armond R. "Toward a Black Media Philosophy Toward a Black Media Philosophy." *Cultural Studies*, published online before print 13 July 2020, doi: 10.1080/09502386.2020.1792524.

Treré, Emiliano. *Hybrid Media Activism: Ecologies, Imaginaries, Algorithms*. London: Routledge, 2018.

Triplett, Norman. "The Psychology of Conjuring Deceptions." *American Journal of Psychology* 11.4 (1900), 439–510.

Trower, Tandy. "Bob and Beyond: A Microsoft Insider Remembers." *Technologizer*, 29 March 2010. Available at https://www.technologizer.com/2010/03/29/bob-and-beyond-a-microsoft-insider-remembers. Retrieved 19 November 2019.

Trudel, Dominique. "L'abandon du projet de construction de la Tour Lumière Cybernétique de La Défense." *Le Temps des médias* 1 (2017), 235–50.

Turing, Alan. "Computing Machinery and Intelligence." *Mind* 59.236 (1950), 433–60.

Turing, Alan. "Lecture on the Automatic Computing Engine" (1947). In *The Essential Turing*, edited by Jack Copeland (Oxford: Oxford University Press, 2004), 394.

Turkle, Sherry. *Alone Together: Why We Expect More from Technology and Less from Each Other*. New York: Basic Books, 2011.

Turkle, Sherry. ed. *Evocative Objects: Things We Think With*. Cambridge, MA: MIT Press, 2007.

Turkle, Sherry. *Life on the Screen: Identity in the Age of the Internet*. New York: Weidenfeld and Nicolson, 1995.

Turkle, Sherry. *Reclaiming Conversation: The Power of Talk in a Digital Age*. London: Penguin, 2015.

Turkle, Sherry. *The Second Self: Computers and the Human Spirit*. Cambridge, MA: MIT Press, 2005.

Turner, Fred. *From Counterculture to Cyberculture: Stewart Brand, the Whole Earth Network, and the Rise of Digital Utopianism*. Chicago: University of Chicago Press, 2006.

"The 24 Funniest Siri Answers That You Can Test with Your Iphone." Justsomething.co, Available at http://justsomething.co/the-24-funniest-siri-answers-that-you-can-test-with-your-iphone/. Retrieved 18 May 2018.

Uttal, William, *Real-Time Computers*. New York: Harper and Row, 1968.

Vaccari, Cristian, and Andrew Chadwick. "Deepfakes and Disinformation: Exploring the Impact of Synthetic Political Video on Deception, Uncertainty, and Trust in News." *Social Media and Society* (forthcoming).

Vaidhyanathan, Siva. *The Googlization of Everything: (And Why We Should Worry)*. Berkeley: University of California Press, 2011.

Vara, Clara Fernández. "The Secret of Monkey Island: Playing between Cultures." In *Well Played 1.0: Video Games, Value and Meaning*, edited by Drew Davidson (Pittsburgh: ETC Press, 2010), 331–52.

Villa-Nicholas, Melissa, and Miriam E. Sweeney. "Designing the 'Good Citizen' through Latina Identity in USCIS's Virtual Assistant 'Emma.'" *Feminist Media Studies* (2019), 1–17.

Vincent, James. "Inside Amazon's $3.5 Million Competition to Make Alexa Chat Like a Human." *Verge*, 13 June 2018. Available at https://www.theverge.com/2018/6/13/17453994/

amazon-alexa-prize-2018-competition-conversational-ai-chatbots. Retrieved 12 January 2020.

Von Hippel, William, and Robert Trivers. "The Evolution and Psychology of Self Deception." *Behavioral and Brain Sciences* 34.1 (2011): 1–16.

Wahrman, Dror. *Mr Collier's Letter Racks: A Tale of Art and Illusion at the Threshold of the Information Age*. Oxford: Oxford University Press, 2012.

Wallace, Richard S. "The Anatomy of A.L.I.C.E." In *Parsing the Turing Test*, edited by Robert Epstein, Gary Roberts, and Grace Beber (Amsterdam: Springer), 181–210.

Walsh, Toby. *Android Dreams: The Past, Present and Future of Artificial Intelligence*. Oxford: Oxford University Press, 2017.

Wardrip-Fruin, Noah. *Expressive Processing: Digital Fictions, Computer Games, and Software Studies*. Cambridge, MA: MIT Press, 2009.

Warner, Jack. "Microsoft Bob Holds Hands with PC Novices, Like It or Not." *Austin American-Statesman*, 29 April 1995, D4.

Warwick, Kevin, and Huma Shah. *Turing's Imitation Game*. Cambridge: Cambridge University Press, 2016.

Watt, William C. "Habitability." *American Documentation* 19.3 (1968), 338–51.

Weil, Peggy. "Seriously Writing SIRI." *Hyperrhiz: New Media Cultures* 11 (2015). Available at http://hyperrhiz.io/hyperrhiz11/essays/seriously-writing-siri.html. Retrieved 29 November 2019.

Weizenbaum, Joseph. *Islands in the Cyberstream: Seeking Havens of Reason in a Programmed Society*. Duluth, MN: Litwin Books, 2015.

Weizenbaum, Joseph. *Computer Power and Human Reason*. New York: Freeman, 1976.

Weizenbaum, Joseph. "Contextual Understanding by Computers." *Communications of the ACM* 10.8 (1967), 474–80.

Weizenbaum, Joseph. "ELIZA: A Computer Program for the Study of Natural Language Communication between Man and Machine." *Communications of the ACM* 9.1 (1966), 36–45.

Weizenbaum, Joseph. "How to Make a Computer Appear Intelligent." *Datamation* 7 (1961): 24–26.

Weizenbaum, Joseph. "Letters: Computer Capabilities." *New York Times*, 21 March 1976, 201.

Weizenbaum, Joseph. "On the Impact of the Computer on Society: How Does One Insult a Machine?" *Science* 176 (1972), 40–42.

Weizenbaum, Joseph. "The Tyranny of Survival: The Need for a Science of Limits." *New York Times*, 3 March 1974, 425.

West, Emily. "Amazon: Surveillance as a Service." *Surveillance & Society* 17 (2019), 27–33.

West, Mark, Rebecca Kraut, and Han Ei Chew, *I'd Blush If I Could: Closing Gender Divides in Digital Skills through Education* UNESCO, 2019.

Whalen, Thomas. "Thom's Participation in the Loebner Competition 1995: Or How I Lost the Contest and Re-Evaluated Humanity." Available at http://hps.elte.hu/~gk/Loebner/story95.htm. Retrieved 27 November 2019.

Whitby, Blay. "Professionalism and AI." *Artificial Intelligence Review* 2 (1988), 133–39.

Whitby, Blay. "Sometimes It's Hard to Be a Robot: A Call for Action on the Ethics of Abusing Artificial Agents." *Interacting with Computers* 20 (2008), 326–33.

Whitby, Blay. "The Turing Test: AI's Biggest Blind Alley?" In *Machines and Thought: The Legacy of Alan Turing*, edited by Peter J. R. Millican and Andy Clark (Oxford: Clarendon Press, 1996), 53–62.
Wiener, Norbert. *Cybernetics, or Control and Communication in the Animal and the Machine*. New York: Wiley, 1948.
Wiener, Norbert. *God & Golem, Inc.: A Comment on Certain Points Where Cybernetics Impinges on Religion*. Cambridge, MA: MIT Press, 1964).
Wiener, Norbert. *The Human Use of Human Beings*. New York: Doubleday, 1954.
Wilf, Eitan. "Toward an Anthropology of Computer-Mediated, Algorithmic Forms of Sociality." *Current Anthropology* 54.6 (2013), 716–39.
Wilford, John Noble. "Computer Is Being Taught to Understand English." *New York Times*, 15 June 1968, 58.
Wilks, Yorick. *Artificial Intelligence: Modern Magic or Dangerous Future*. London: Icon Books, 2019.
Willson, Michele. "The Politics of Social Filtering." *Convergence* 20.2 (2014), 218–32.
Williams, Andrew. *History of Digital Games*. London: Routledge, 2017.
Williams, Raymond. *Television: Technology and Cultural Form*. London: Fontana, 1974.
Wilner, Adriana, Tania Pereira Christopoulos, Mario Aquino Alves, and Paulo C. Vaz Guimarães. "The Death of Steve Jobs: How the Media Design Fortune from Misfortune." *Culture and Organization* 20.5 (2014), 430–49.
Winograd, Terry. "A Language/Action Perspective on the Design of Cooperative Work." *Human–Computer Interaction* 3.1 (1987), 3–30.
Winograd, Terry. "What Does It Mean to Understand Language?" *Cognitive Science* 4.3 (1980), 209–41.
Woods, Heather Suzanne. "Asking More of Siri and Alexa: Feminine Persona in Service of Surveillance Capitalism." *Critical Studies in Media Communication* 35.4 (2018), 334–49.
Woodward, Kathleen. "A Feeling for the Cyborg." In *Data Made Flesh: Embodying Information*, edited by Robert Mitchell and Phillip Thurtle (New York: Routledge, 2004), 181–97.
Wrathall, Mark A. *Heidegger and Unconcealment: Truth, Language, and History*. Cambridge: Cambridge University Press, 2010.
Wrathall, Mark A. "On the 'Existential Positivity of Our Ability to Be Deceived.'" In *The Philosophy of Deception*, edited by Clancy W. Martin (Oxford: Oxford University Press, 2009), 67–81.
Wünderlich, Nancy V., and Stefanie Paluch. "A Nice and Friendly Chat with a Bot: User Perceptions of AI-Based Service Agents." *ICIS 2017: Transforming Society with Digital Innovation* (2018), 1–11.
Xu, Kun. "First Encounter with Robot Alpha: How Individual Differences Interact with Vocal and Kinetic Cues in Users' Social Responses." *New Media & Society* 21.11–12 (2019), 2522–47.
Yannakakis, Georgios N., and Julian Togelius. *Artificial Intelligence and Games*. Cham, Switzerland: Springer, 2018.
Young, Liam. '"I'm a Cloud of Infinitesimal Data Computation': When Machines Talk Back: An Interview with Deborah Harrison, One of the Personality Designers of Microsoft's Cortana AI." *Architectural Design* 89.1 (2019), 112–17.

Young, Miriama. *Singing the Body Electric: The Human Voice and Sound Technology*. London: Routledge, 2016.
Zdenek, Sean. "Artificial Intelligence as a Discursive Practice: The Case of Embodied Software Agent Systems." *AI & Society* 17 (2003), 353.
Zdenek, Sean. "'Just Roll Your Mouse over Me': Designing Virtual Women for Customer Service on the Web." *Technical Communication Quarterly* 16.4 (2007), 397–430.
Zdenek, Sean. "Rising Up from the MUD: Inscribing Gender in Software Design." *Discourse & Society* 10.3 (1999), 381.

INDEX

For the benefit of digital users, indexed terms that span two pages (e.g., 52–53) may, on occasion, appear on only one of those pages.

actor-network theory, 9–10
advertising, 124–25, 128
affordances, 8
agency, 9–10, 14, 72, 131
Alexa, 24, 29–30, 61–62, 84, 103–4, 107–25
algorithmic imaginary, 65–66
algorithms, 64–66, 109–10, 122
Amazon, 112, 122–23
anecdotes, 57–58
animals, 27
anthropomorphization, 64, 114
Appadurai, Arjun, 14, 72
Apple, 47–48, 61–62, 107–8, 112, 122–23
art, 4, 97–98
artifacts, 49, 51, 61–62, 65–66
artificial intelligence (AI)
  AI and culture, 21–22
  AI as discipline, iv, 18–19, 31, 52, 66
  AI as medium, 32
  AI as myth, 7–8, 17, 18–19, 31, 34–38, 41, 43, 45, 48–49, 51–52, 56, 58–61, 70, 91–92, 104, 115–17
  AI companions, 65, 118
  AI Winter, 68–69, 89
  communicative AI, 8, 11, 105, 124, 128
  credibility of AI, 20
  criticism against AI, 65, 68–69, 91–92
  definition of AI, 3–4, 27–28, 36, 52, 127
  GOFAI (Good Old-Fashioned Artificial Intelligence), 33–38
  "golden age" of AI, 33–38
  "strong" AI, 3–4, 34–38
  symbolic AI, 35–38
artificial life, 43–44
automata, 12–13, 91
avatars, 62

Bell Labs, 43–44
biographies of media, 72–74
black box, 11, 33, 36–37, 47–48, 64, 65–66
Bletchley Park, 28, 87
bots, 73–74
Bourdieu, Pierre, 11, 98–99
brain-computer analogy, 18–19, 48–49

CAPTCHA, 104–5
characterization, 61–62, 78, 80–81, 100, 101, 102–3, 111, 113–15, 117–18
chatbots, 21–22, 24, 50, 75–76, 87, 115–17, 129
chess, 12–13, 25–27, 55–56, 74–75, 89–90
children, 129–30
cinema, 12, 34–35, 38, 39, 111
Clarke, Arthur C., 38
class, 10, 42, 102, 103, 113, 129–30
cloud, 123, 124–25
Colby, Kenneth Mark, 59, 102–3
communication, 22–23, 31, 74
  communication theory, 23–24, 41, 128
  Computer-Mediated Communication, 23, 96

Computer History Museum, California (previously in Boston, Massachusetts), 89–90
computing
computer games (*see* digital games)
computer history, 3, 9–10, 25, 69–70
computer industry, 47, 48, 69, 80, 108, 122
computer networks, 23
Computers Are Social Actors (CASA) paradigm, 14–15, 82–83
computer science (discipline), 3, 18, 39–40
computer skills, 45–46, 47–48, 80
computer users (*see* user)
conjuring. *See* magic
consciousness, 3, 18–19, 36
controversies
around AI, 2, 51, 59, 63–64, 91–92, 127
around the Turing Test, 20, 87–88
conversation, 1, 53–54, 55, 75, 76–78, 92–93, 99, 100, 102, 105–6, 115–20
CONVERSE (chatbot), 95
Corbato, Fernando, 71, 73–74
Crowther, Will, 79
cryptography, 28–29
cybernetics, 18, 40–41
cyborg, 24

daemons (computing), 70–74, 85
Dartmouth Conference, 18, 33
data, 53–54, 111–12, 117–18, 120, 130
deception
banal deception, 7–11, 30, 39, 45–46, 48–49, 65, 79, 105, 108, 114, 120, 124, 125, 128–32
playful deception, 29–30, 47
self-deception, 30
"straight-up" deception, 1, 7, 12–13, 105, 120, 124–25, 128–29
wilful deception, 30, 32, 47–48, 64–65
decision trees, 76–77
Deep Blue, 55–56
deepfakes, 128–29
deep learning, 3, 18, 111–12, 115–17, 130
design, 30
de Vaucanson, Jacques, 12–13
dialogue trees, 74–79, 85
digital games, 19, 26–27, 55–56, 61, 74–79, 85
digital revolution, 70
disembodiment, 11, 23–24, 40–41, 115
DOCTOR (chatbot), 54. *See also* ELIZA
Dreyfus, Herbert, 68–69
drosophila, 25–26

Edison, Thomas Alva, 110–11
electronic brain, 19, 24, 45, 58
ELIZA (chatbot), 49, 50–67, 69–70, 78, 79, 101–2
Eliza effect, 51–52, 64–65, 67, 78–79
email, 70–71, 120–21
emotions, 6, 105–6, 118, 124–25
entertainment, 7, 29–30
Epstein, Robert, 87, 89, 90, 91, 94, 96–97
ethics, 15, 59, 60, 130–31
extensions, 24, 38–39, 102

Faraday, Michael, 16–17, 20, 22
feedback mechanisms, 40–41
filter bubble, 121
Fox sisters, 16
future, 13, 37–38, 91–92

games, 25–27, 29, 32, 43–44. *See also* digital games
Gell, Alfred, 14, 72
gender, 10, 29, 42, 102–4, 112–14, 129–30, 131–32
genetics, 25–26
Gombrich, Ernst, 4, 97–98
Google
Google Assistant, 107–25
Google (company), 1, 112, 115, 122–23
Google Duplex, 1–2, 24
Google Search, 120–21, 122
Greenberger, Martin, 41, 46

habitus, 11, 98–99
hackers, 26
Hagelbarger, David, 43–44
haiku, 117
HAL (fictional computer) 59
Hayles, Katherine, 40
history of science, 25–26, 44, 95
history of technology, 55–56, 80
Hofstadter, Douglas, 37

Huizinga, Johan, 28–29
human
human bias, 10, 20–21, 22, 42, 91–92, 102–4, 113, 130
human body, 23–24
human brain, 36–37, 40–41
human communication, 3, 92, 93, 97
human senses, 5–6, 12, 26–27, 110–11, 112
human vs computer, 26, 27, 28–29, 36–37, 40, 55–56, 74–75, 76–77, 89–91, 104, 129–30
Human-Computer Interaction (HCI), 4–5, 7–8, 9–10, 14, 23–24, 26–28, 31, 34, 38–44, 53, 66, 71–72, 74–75, 98, 104–5, 112, 123–24
human-computer symbiosis, 40–41, 66, 75
Human-Machine Communication (HMC), 11–12, 14, 20–21, 23–25, 31, 40–41, 55, 75, 87, 95–96, 97, 104–5, 107
Humphrys, Mark, 95–96
Hutchens, Jason, 95

IBM, 47–48, 55–56
imagination, 9–10, 43–44, 46, 60–61, 113–14, 123–24
imitation game. *See* Turing test
information retrieval, 109–10, 120–23
information theory, 23–24, 40–41
infrastructure, 71–72
intelligence, definition of, 3–4, 15, 18–19, 23, 27, 31, 36, 42–43, 115
interdisciplinarity, 35–36
interface
conceptualizations of interface, 2–3, 10, 23–25, 26–27, 39, 45–48, 53, 79, 108, 109, 121
graphic user interfaces (GUI), 46
seamless interfaces, 47, 48, 71–72, 81–82, 118
social interfaces, 79–84, 85, 103–4, 109
voice-based interfaces, 110–11
web interfaces, 120–23, 124–25
internet, 22, 62, 70, 74, 96, 120–23. *See also* World Wide Web
intuition, 36
irony, 61–62, 95, 117–18, 122–23

Jackson, Samuel Lee, 112
Jibo (robot), 6, 118
Joe the Janitor (chatbot), 100–1
journalism, 36, 37, 48–49, 58, 60, 81–83, 91
Julia (chatbot), 102–3

Kronrod, Alexander, 25–26
Kubrick, Stanley, 38, 59

labor, 112, 122–23
Lady's Oracle (game), 29
language, 65–66, 76–77, 92, 94–95, 97–99, 111, 118–20
Lanier, Jaron, 64–65
Latour, Bruno, 9–10, 57–58, 91–92
Laurel, Brenda, 26
Licklider, J. R. C., 40, 41
Lighthill report, 68–69
Lippmann, Walter, 113–14
literary fiction, 45–46, 57–58, 62, 78, 100, 102–3
Loebner, Hugh, 87, 89
Loebner Prize, 23, 27–28, 87, 117
Lovelace, Ada, 70
Lovelace objection, 70
lying, 27–28

machine learning, 18, 111–12
magic, 29, 34–35, 45, 46–49, 58–59
Massachusetts Institute of Technology (MIT) 37, 38, 41, 50, 70, 72–73
mathematics, 18
Maxwell, James Clerk, 71, 73
May, Theresa, 92
McCarthy, John, 33, 45, 47–48
McCulloch, Warren, 18
McLuhan, Marshall, 24, 30, 34–35, 102, 114–15, 128–29
Mechanical Turk (automaton chess player), 12–13, 91
media archaeology, 13
media history, 11–13, 22, 23–24, 28, 32, 34–35, 39, 45–46, 97, 110–11, 124–25
media theory, 9, 24–25, 30, 34–35, 46, 102, 114–15, 128–29
mediation, 22, 24–25, 97, 110–11, 124
medium, concept of, 11–12, 24–25, 71, 124

metaphors, 8, 36, 46, 55, 60–61, 123–24
MGonz (chatbot), 95–96
Microsoft
  Microsoft Bob, 79–84, 86
  Microsoft (company), 47–48, 79–84, 101, 103–4
  Microsoft Cortana, 107–8, 122–23
  Microsoft Publisher, 83–84
  Ms. Dewey, 103–4
mindlessness, 8–9, 83–84
mind reading, 43–44, 117–18
Minsky, Marvin, 33, 36, 37, 38, 42–43
mobile media, 107–8, 114–15
Münsterberg, Hugo, 95
mythology, 30, 71, 72–73

Nagel, Thomas, 18–19
Narcissus, myth of, 30
narratives about technology, 56–61, 63–64, 91–92
Nass, Clifford, 14–15, 82–83
natural language processing, 50–51, 69–70, 107–8, 109–10, 115–20
Negroponte, Nicholas, 89
Netanyahu, Benjamin, 129
network. *See* computer networks
neural networks, 3, 18, 111–12, 130
Newell, Allen, 33
New York Times, 91

observer, iv, 19, 27–28
Oettinger, Anthony G., 47
opacity. *See* black box
optical media, 46, 51–52, 57

PARRY (chatbot), 59, 101–2
Pask, Gordon, 43–44
PC Therapist (chatbot), 94, 99
personal computing, 42, 69, 79–84, 89
Peters, John Durham, 22, 122–23
philosophical toys, 51–52, 57, 62
philosophy, 5, 18–19
phonetics, 54–55, 112
Photoshop (software), 109
physics, 44, 71, 73
Pichai, Sundar, 1
Pirate Bay, 74
Pitts, Walter, 18
play and playfulness, 8–9, 25, 28–30, 32, 57, 62, 78–79, 85
politics, 92, 124–25, 128, 129
programming, 4–5, 9–10, 45–46, 52–53, 56, 70, 71, 75–76, 77, 102, 118–20, 131
programming languages, 35, 76–77, 118–20
psychology, 5–6, 8–9, 12, 18, 21–22, 30, 37–54, 59, 61–62, 112
psychotherapy, 54–55, 62
psychotherapy chatbots, 21–22, 59, 65
Pygmalion, 54–55

quantum mechanics, 44

race, 10, 42, 102, 103, 113, 129–30
Reeves, Byron, 14–15, 82–83
Replika, 6, 118
robot apocalypse, 15, 28–29, 38
robots, 9, 11, 24, 118, 128–29
Romero, George, 27, 30

Samuel, Arthur, 37–38
science fiction, 38, 48–49, 58, 59, 104
scripts, 9–10, 53–54
search engines, 120–22
Secret of Monkey Island (digital game), 76
semiotics, 100
Shannon, Claude, 18, 25, 33, 43–44
Shaw, George Bernard, 54–55
Shieber, Stuart, 91, 94
Simon, Herbert, 33, 41
simulation, 105–6, 118
Siri, 24, 29–30, 50–51, 61–62, 107–25
smartphones, 7–8, 24, 114–15
smart speakers, 107–8, 118
social interaction, 30
social life of things, 14, 51, 72–73
social media, 22, 24, 101, 104–5, 110–11, 121
social media bots, 24, 101, 105, 125, 129
social networks. *See* social media
software, 4–5, 48, 51–52, 55–56, 61, 67, 68, 109–10
sound recording, 12, 39, 97
spectacle, 90–91, 104
speech dialog systems. *See* voice assistants
speech processing, 109–15
spiritualism, 16–17, 21, 22, 29, 34–35, 58–59, 95
Star Wars, 84

stereotyping, 102–4, 113–14, 128, 129–30
storytelling, 57–58, 100, 101
Strimpel, Oliver, 89–90
Suchman, Lucy, 14, 40–41, 68, 84–85, 97
superstition, 58–59
surveillance, 112
suspension of disbelief, 100

Tay (social media bot), 101
technology, 3, 7–8, 24
  technological failure, 80–83
  technological imaginary, 34–38, 55–56, 115–17
techno-utopia, 63–64
telegraph, 23–24
television, 45–46
theatre, 54–55, 89, 100, 117
theology, 108–9
time-sharing, 23, 26, 41–43, 46, 57, 68, 72–73
TIPS (chatbot), 99, 100
transparent computing, 45–49, 71–72, 83–84, 120, 130–31
trickery, 27–28
trinity, 108–9
Turing, Alan, 2–3, 16, 44, 52, 74–75, 89, 115, 127
Turing test, 2–3, 16, 44, 49, 62–63, 74–75, 76–77, 87
Turkle, Sherry, 14–15, 21–22, 57, 64, 65, 69, 78–79, 96, 104–5
Twitter, 24, 121
typewriter, 23–24

Union of Soviet Socialist Republics (USSR), 25–26
United States of America, 16
user, 3, 7–8, 10, 23, 26, 27–28, 41–43, 46, 47–48, 53, 68, 79, 117–18, 124, 131–32
user-friendliness, 8–9, 47–48, 81–82, 120, 130–31

Victorian age, 29
videogames. *See* digital games
vision, 97–98
voice, 107–8, 110–15
voice assistants, 2–3, 7–8, 9, 24, 29–30, 61–62, 65, 73–74, 77, 84–85, 103–4, 105, 107, 129–30

war, 28–29
web. *See* World Wide Web
Weintraub, Joseph, 94, 102–3
Weizenbaum, Joseph, 25, 42–43, 50, 78, 79, 130–31
Welles, Orson, 12
Whalen, Thomas, 99–100
Wiener, Norbert, 18, 27
Wired (magazine), 89
World Wide Web, 70–71, 95, 103–5, 109–10, 120–21, 123, 124

Youtube, 84

Zork (digital game) 78–79

Printed in the USA/Agawam, MA
September 12, 2022

798393.061